MYSTERIES OF
THE HUMAN BODY

Cover: A ''topographical map'' of the body's complex contours, rendered by two cameras, is displayed against a supermagnified cross-section of alveoli, air cells of the lungs.

MYSTERIES OF THE HUMAN BODY

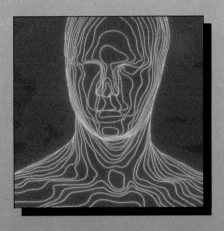

By the Editors of Time-Life Books

TIME-LIFE BOOKS, ALEXANDRIA, VIRGINIA

CONTENTS

A LIFE-WORTHY VESSEL

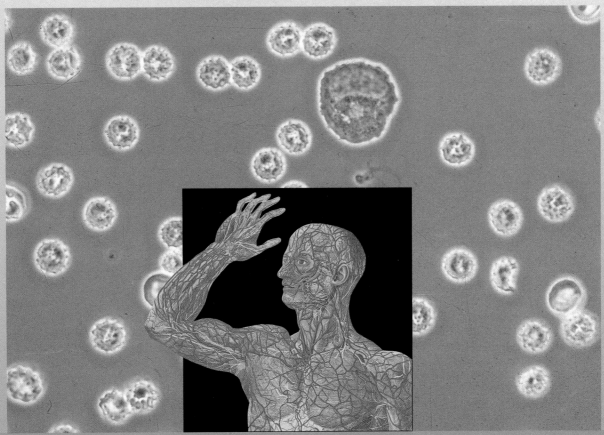

The human body, which American poet Walt Whitman called the ship of the soul, is an eminently practical vessel, crammed with a bewildering labyrinth of pumps, valves, and plumbing, all managed by an electrochemical signals network beyond the wildest dreams of engineers. Alternatively, the body can be viewed as a kind of metropolis of smaller creatures. Trillions of cells live in a superbly run community, cooperating for the general good even as they pursue their own specialized roles—whether it be the detection of light or sound, the manufacture of substances that help digest food, or the fabrication of defenses against viral invaders.

Outside, as though above the endless chemical stirrings within, stands the total, physical self that we ourselves are able to see. Strung upon a flexible framework of bone and tissue that is animated by the counterplay of fleshy machines called muscles, the body possesses spectacular abilities. It is supple and swift, powerful, resourceful, constantly renewed —but also mortal.

Ironically, knowledge of death, like understanding of the body itself, springs from a quick intelligence that somehow lives in a ruling inner organ. Rich with understanding of the world, life, time, and mortality, human consciousness is nevertheless a prisoner of the brain, doomed to share the fate of the larger vessel in which both must reside.

A Baffling Machine

"I tremble lest I have mankind at large for my enemies," said William Harvey as he began a lecture to London's Royal College of Physicians in 1616. A cautionary note was in order; minutes later, the small, round-faced doctor demonstrated how blood moves from the heart in a closed loop through the body—and discredited medical wisdom that had endured for some fifteen hundred years.

The prevailing medical description of the blood's circulation was the legacy of Galen, a second-century Greek physician. He believed that new blood was constantly being created in the liver from a milky fluid he called chyle and flowed out through the veins, bearing "Natural Spirits" to the tissues. Some of this blood was drawn up by the heart, mixed with air, and sent through the arteries, carrying livelier "Vital Spirits" to the tissues and the brain. As a surgeon to gladiators, he had seen much of the body's interior, and his explanations of how blood behaved carried great weight.

Even in Harvey's day, anatomies were said to be abnormal if they did not match Galen's model, still the starting point for any circulation theory. Thus, although doctors knew that blood moved, they saw the movement as a random ebb and flow, like changing drafts in a room. They also clung to Galen's belief that two kinds of blood flowed in the body, one to the right side of the heart, the other to the left. Within the heart, according to Galen, blood passed through the septum—the vertical wall that divides the organ in half. The heartbeat signified a dilation by the spirits borne upon the blood, not a muscular spasm.

Despite centuries of observation, however, no one had made the great leap of understanding attributed to Harvey, then in his thirty-eighth year. Although a staunch Galenist himself, he approached his subject through a series of careful experiments. He conducted a number of human dissections and also observed the exposed living hearts of vivisected animals. In addition, Harvey examined reptiles and amphibians, whose cold-bloodedness slowed down the heartbeat to a sluggish, rhythmic contraction. "I began to think whether there might not be a movement, as it were, in a circle," he wrote in his journal in 1615. A year later, he dropped his bombshell at the Royal College.

The blood, he explained to the assembled physicians, moves in a single direction, out from the left side of the heart through the arteries, returning to the right side of the heart by way of veins. A series of one-way valves prevents it from washing back toward its source. The septum is impermeable, the blood moving from the right side to the left side via a circulation through the lungs.

There is only one kind of blood, he said, constantly recirculated through the body, propelled by the contraction of the heart, where circulation begins and ends. The pulse was the pressure wave transmitted as a surge of blood shot out of the contracting heart.

Harvey bolstered his argument with simple logic: Hundreds of pounds of blood pass through the heart every hour; if it is not being recirculated, where does it come from and where does it go? His ideas, not published until twelve years later, were initially attacked by Galenists but quickly became medical dogma.

They also became the stuff of mystery for medical historians. Why, they wondered, did Harvey go straight to the idea of a circulation, when such early masters as Leonardo da Vinci had circled the same idea without seeing it? Galen had used cooking, brewing, and smelting analogies to show how the body functioned. Perhaps earlier researchers had been blinded to the heart's true function by the absence of a suitable technical analogy. But, by Harvey's time, mechanical pumps—developed to draw seepage out of ever-deepening mines—had become common in fountains and firefighting gear. Some scholars suggest that the pulsing jet of water from a fire brigade's handworked pump (above) was probably the crude metaphor in which Harvey recognized the human heart as a tireless blood-pumping machine. □

The Missing Link

In the circulatory system described by William Harvey, blood made its way from arteries to veins by unknown means. While his closed-loop circulation theory required some kind of connecting passages, he was unable to detect them with the naked eye. But in 1661, four years after Harvey's death, the Italian physician Marcello Malpighi used the newly invented microscope to find the vital connectors: tiny vessels called capillaries.

These delicate conduits spread through tissue as networks of frail, thin-walled channels only a few hundredths of an inch long. The smallest capillaries are so narrow that red blood cells, which are only about three ten-thousandths of an inch in diameter, must form a single file as they pass through, bending and squeezing their way along the narrow passage *(below)*. As the blood makes its way through these straits, it yields its life-giving cargo of oxygen and food to adjacent tissue, picks up cellular wastes, and finally enters the veins, bound once more for the heart and the lungs. □

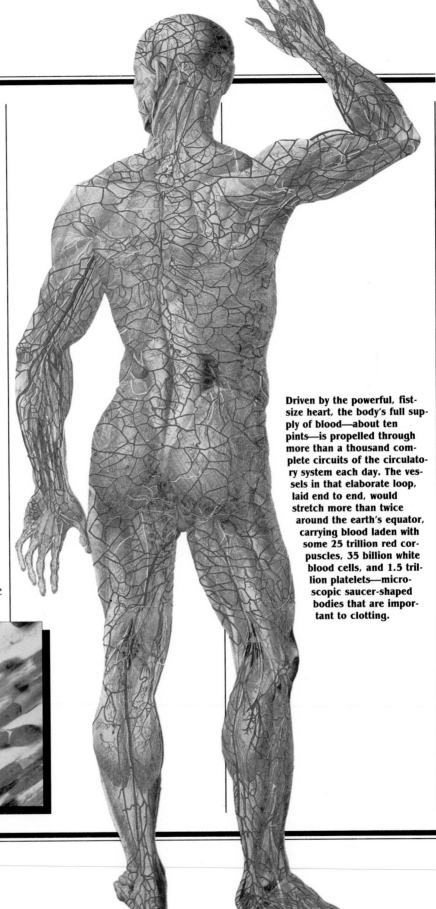

Driven by the powerful, fist-size heart, the body's full supply of blood—about ten pints—is propelled through more than a thousand complete circuits of the circulatory system each day. The vessels in that elaborate loop, laid end to end, would stretch more than twice around the earth's equator, carrying blood laden with some 25 trillion red corpuscles, 35 billion white blood cells, and 1.5 trillion platelets—microscopic saucer-shaped bodies that are important to clotting.

Strange Bloodfellows

In Europe, the first successful blood transfusion was carried out in 1665 by a Cornish physician named Richard Lower, who used hollow goose quills to pipe arterial blood from one dog into the veins of another. Two years later, French physician Jean-Baptiste Denis took the bold step of shunting eight ounces of sheep's blood into the veins of a human. The patient's survival led Denis to try his technique on others, including Antoine Mauroy, who survived two transfusions of calf's blood but died following a third attempt. His widow brought a charge of murder. Although Denis escaped conviction, the trial led to a ban on all blood transfusions in France. Within a few years, the practice was outlawed in Italy and abandoned by doctors in England. Then, for two hundred years, it largely vanished from the medical scene.

But a latent interest in transfusion as a therapeutic device remained. In the early nineteenth century, London obstetrician James Blundell became interested in the technique as a way of saving the many women who died from hemorrhage during childbirth. Preliminary experiments with dogs demonstrated that, even when severely bled, they could be resuscitated by an immediate transfusion of blood from another dog. At the same time, Blundell warned others away from mixing the blood of different species, noting that it held "great danger, and in general death ensues." Working with moribund humans, he began cautiously transfusing their blood. In 1828, the British medical journal *Lancet* reported that Blundell had performed an apparently successful transfusion on a human. Others, less conservative, resumed experiments blending animal and human blood, which were invariably fatal unless small amounts of animal blood were used. But even human-to-human transfusions were perilously uncertain. In some cases, patients flourished after receiving another's blood; usually, they died.

In 1900, the unpredictability of human-to-human transfusions was explained. Vienna-born Karl Landsteiner observed that when blood from two people was mixed, the red cells often bunched together, signaling incompatibility. Although the detailed workings of the body's immune system were then unknown, he theorized correctly that this clumping is caused when an agent within the red cells (later identified as an antigen) reacts with another agent in plasma, the liquid medium in which blood cells are borne. By sorting blood mixtures according to the way red cells behaved, he identified four major groups: one containing agent A, another containing B, a third containing both (AB), and a fourth containing neither (O).

Like can safely be mixed with like, and type O blood, which contains none of the clot-causing antigens, can be added to the other three without reaction. But O, the universal donor, cannot receive from others; it reacts to antigens in other blood types, causing a fatal clotting. Landsteiner's discovery led to a resurgence of therapeutic blood transfusion and to a Nobel prize in 1930. □

Poison Control

A book-size clay crescent displayed at London's British Museum is scored with a pattern of small squares, like the grid on a map *(right)*. The figure was discovered in about 1890, an artifact of the Babylonian culture of the eighteenth or nineteenth century BC. It is indeed a chart, but not of land. Instead, it seems to have been designed to help haruspices—readers of entrails—find their way around a sheep's liver, whose texture and terrain, it was once believed, revealed the future.

The liver, which occurs only in vertebrates, has the look of life's major organ. At about three pounds, it is the human body's largest gland and boasts a dual blood supply. Veins feed into the thick-skinned, maroon vessel en route from the intestines to the heart, while arteries bring fresh blood from the heart to the liver. Seeing this seeming confluence, the ancient societies of Babylon, Mesopotamia, Greece, and Rome took the liver to be the body's physical and spiritual center.

It is, in fact, a remarkable organ. On the slick surface, its design is not much more complicated than that of the Babylonian map. Viewed microscopically, however, the liver becomes a three-dimensional labyrinth of trapezoidal rooms built of connected hexagonal cells, like floor tiles, elaborately tunneled with veins, arteries, capillaries, and minuscule ducts for transporting bile, a bitter, yellow-green fluid essential to digestion. Like a compact chemical plant, the liver filters blood returning to the heart from the intestines. It stores sugar, vita-

mins, and other foods, and produces such cellular building blocks as proteins and amino acids. But it does its hardest work, and takes its gravest risks, as the body's poison control center.

The heart-bound veins passing through the liver are filled with more than blood; they have also taken on a cargo of material from the world outside the body. Substances that are as various as arsenic and gold, mercury and TNT, carbon tetrachloride, and snake venom can all be filtered out of the bloodstream in the tunnels of the human liver.

The worst enemy of the liver, however, is not those substances, but alcohol, a deadly poison to the organ that must oxidize it. Too much, over too long a period of time, produces the effects seen in cirrhosis: The liver becomes engorged with yellow fat, swelling against the rib cage, its exterior showing a lumpy hardening and scars. As continued alcohol consumption propels the disease further, the labyrinthine channels within are torn and twisted, and finally collapse. Liver function diminishes. The swollen body drifts toward debility and death.

Incredibly, the liver may function until nine tenths of its substance has been destroyed. And if, at any point along this grim downhill road, the flow of alcohol is stopped, the organ quickly rebuilds itself. Like a starfish regenerating a lost arm, the liver casts off its fatty yellow shroud and most of the symptoms that go with it, restores the damaged labyrinth to functional condition, and shrinks to its normal size. □

The Brain's Inner Universe

Like a flower on a bony stem, the human brain *(right)* blooms at the top of the spine, sheathed by a leathery cover inside the protective helmet of the skull. It contains an estimated hundred billion neurons, the nerve cells that propel the conceptual, sensory, and motor functions of the body.

Santiago Ramón y Cajal, a pioneering Spanish neuroscientist of the nineteenth century, called the neuron "the aristocrat among the structures of the body." Like an aristocracy, they are grand users of resources. Although the brain accounts for only 2 percent of body weight, it bathes in about 15 percent of the blood at any given instant and consumes 20 percent of the food and oxygen.

Each neuron has a central nucleus from which branching extensions—dendrites—reach out, along with a lengthened appendage called the axon, which branches like a plant root at its far end. The dendrites, several hundred thousand miles of them per human body, are the listening antennas of the cells; the axons conduct outgoing signals. In the linkage between neurons, the axon of one and the dendrite of another almost touch, but not quite. They are chemically connected across a millionth of an inch wide gap known as the synapse. There are perhaps a hundred trillion synapses in the whole brain.

The signals that flash around the brain from neuron to neuron begin as electrical impulses discharged, or fired, into the axon. At the synapse, the electrical impulse causes the production of a chemical called a neurotransmitter, which flows across the gap. Arriving at specialized receptors in the other cell's dendrite, the signal is converted back into electricity and continues on its way.

Although the electricity of thought is of low wattage, it is sufficient to overwhelm delicate cerebral circuitry. If a large fraction of the brain's neurons fired at once, the organ would be swept by a crippling electrical storm resembling an epileptic fit. As a defense against its own power, the brain contains strong inhibitors that limit the firing of neurons.

Despite the seeming swiftness of thought, the brain's impulses creep compared to those of man-made electronic systems. The largest nerve fibers carry signals at only about 200 miles per hour; others transmit signals slower than 55 miles per hour, a leisurely pace compared to electrons moving at tens of thousands of miles per second. Scientists explain this relatively slothful transmission rate as the inherent sluggishness of electrochemical processes compared to electronic ones. But some researchers believe it is also an environmental adaptation: The brain must gear its signals to the body's capacity to respond.

In the human fetus, the brain begins to develop after the third week following conception, arising from a sheet of just 125,000 cells—the neural plate. Thereafter, for the remaining eight months of gestation, 250,000 new neurons appear each minute. Some of them stay where they form, but others migrate to a final position that, for objects that are only a few millionths of an inch across, can be a great distance away.

At birth, the brain weighs only about three quarters of a pound. It grows to about one and a half pounds by age one. But neurons cease dividing at birth and do not regenerate themselves as they die away. In a sense, one spends a lifetime learning to do more with progressively fewer neurons.

Scientists have determined that rich and varied experience causes a greater complexity of synaptic connections, and a deprived existence causes a corresponding simplification. But all normal brains appear to be physically similar, with no discernible differences between the cerebral structure of an Einstein and that of an average thinker. □

A Gland for All Seasons

Where the spinal column and the head meet, in a kind of no man's land between the body's automatic reflexes and the cerebral apparatus of thought, a tiny gland called the hypothalamus holds sway.

Although it represents only about one three-hundredths of the brain's volume and weighs only four grams—about nine thousandths of a pound—this potent gland wields an enormous amount of managerial power.

The hypothalamus regulates the body's temperature and water balance, manages the autonomic, or reflex, nervous system, and controls fatigue and hunger. It pro-vides the twenty-eight neurons that are home for the body's physical clock—the suprachiasmatic nucleus, or SCN *(page 114)*. It also operates the brain's elaborate reward-and-punishment system, responding on behalf of the body to pain and pleasure.

So fragile that the gentle touch of a surgeon's sponge can plunge its owner forever into coma, the hypothalamus can nevertheless be probed. Electrically stimulate one part, and the subject responds with rabid aggression; stimulate another, the subject fawns. But the minuscule organ acquires its greatest powers by managing other glands that control the functions of memory, emotional behavior, and sexual reproduction. □

The Painless Zone

Although observers may wince at the sight of surgeons pricking at the brain, where the body sends its pain signals to be read and acted upon, the brain itself feels nothing. The shocks of scalpel or cauterizing pencil or blunt dissector produce no sensation at all. Even the splitting headaches that are caused by tumors and blood clots in the brain originate in the dura mater, a leathery shroud that adjoins the skull. The brain swells and presses against the sheath, which, unlike brain tissue itself, has pain-sensing nerves. □

As the Brain Sees It

The brain perceives the body differently from the way we see ourselves, allocating cerebral space to those parts most in need of coordination. Thus, in terms of their share of brainpower, hands—especially thumbs—shoulders, lips, tongue, and feet loom large. The rest of the body shrinks proportionately in the brain's view, deformed into an assemblage of enlarged or stunted parts, as shown at right. Sight, though, is not represented proportionately here because it has a large section of the brain—the visual cortex—dedicated solely to its service. □

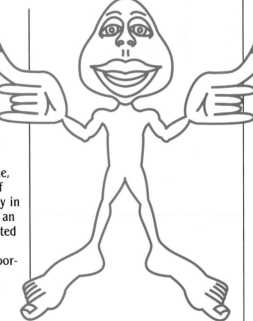

A Multitude of Bodies

The body only seems to be a construct of bones, flesh, and organs; in fact, it is more like a nation of a hundred trillion or so individual cells. Each cell, in turn, contains still smaller living units called organelles, busily building the fundamental chemical stuff of life according to detailed instructions packed into the cell's genes.

From the nucleus, which is about a tenth of the cell's diameter and the brains behind the tiny chemical factory, orders go out that organize the apparatus of pro-duction. Within the nucleus coil forty-six tightly packed threads of deoxyribonucleic acid—DNA—an inch or two long, tens of thousands of times longer than the diameter of the nucleus. On these, drawn from some three billion possible chemical combinations, fifty thousand genes constitute a blueprint for reproducing any one of the body's two-hundred-odd different types of specialized cells.

A still-mysterious object within the nucleus, the nucleolus, has a DNA thread of its own and builds the parts that later join to form the spherical protein-making sites called ribosomes. About a millionth of an inch in diameter, thousands of ribosomes live on a wrinkled sheath around the nucleus. There, they receive DNA instructions and assemble the specified types of protein at a prodigious rate: Each second, more than a million chemical reactions in a cell's ribosomes produce an estimated two thousand protein molecules.

Organelles known as mitochondria, with their own DNA and ribosomes and the ability to reproduce

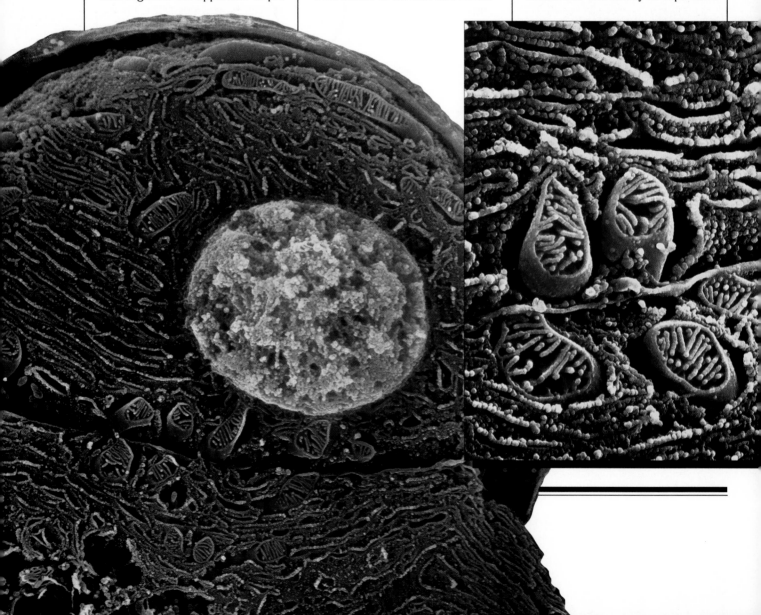

by division, power the cell. Drifting in a watery medium by the hundreds or thousands, mitochondria consume oxygen and convert nutrients into usable energy.

Smaller organelles called lysosomes serve as stomachs in the cell, further digesting foods. Like the full-size stomach, these microscopic guts employ enzymes that would digest the cell itself were they not sealed in a protective membrane. The cell, too, exists within a wall, thick enough to keep its world intact, but, at only sixty millionths of an inch, thin enough to permit the intake of fresh supplies and the ejection of wastes.

Within the bounds of that membrane, the body's cells vary widely in size and behavior. Many are only about one twenty-five hundredths of an inch across, although certain gangly nerve cells may be four feet long. Some are created and die in the same place, tied eternally to their adjacent comrades. Others, such as the soldiers of the immune system, patrol on their own. And sperm cells, during their brief life, are as mobile as a school of tadpoles.

Such independence goes beyond what must be done for the body's collective good. Many cells, removed from their host, can live as independent organisms. In a microaquarium of nutrient-rich, oxygenated water, a detached cell will trade the form it took as a specialist in the body for that of an ancient ancestor, the one-celled protozoan. The amoebic blob crawls about its new home, foraging like a hungry blood cell and reproducing by division. Even after millions of years' labor in the huge collective of the human body, a cell recalls how to survive alone. □

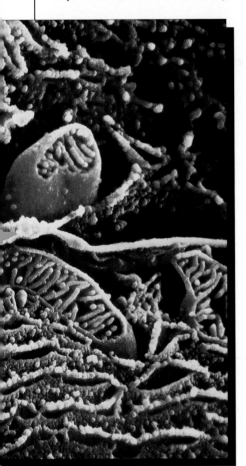

The protective membrane surrounding a cell *(far left),* shown in green, has been peeled back to reveal the complex inner structure. The light yellow nucleus contains bundled chromosomes that carry the cell's genetic blueprint. A magnified detail *(near left)* shows the saclike mitochondria that convert nutrients into energy, surrounded by legions of small, round, protein-making ribosomes.

"Small" Packages

The so-called small intestine *(above)* consists of some twenty-two feet of inch and a half tubing, coiled between the bottom of the stomach and the top of the wider, six-foot-long large intestine just below. A model of natural packaging, the small intestine carries much of the burden of digestion. In fact, it can sustain life without the help of esophagus, stomach, or large intestine. However, this vital organ's power does not come from its great length, but from its inner design: The folds and hairlike feelers—or villi—lining the small intestine's interior provide a nutrient-absorbing area of about 2,700 square feet, roughly the size of a tennis court. □

Skinscape

Although the body encloses most of its organs, it is itself wrapped in the largest one, the skin. Stripped away and spread, the skin of an average adult male would cover about twenty square feet, a female's about seventeen.

Skin constantly flakes away, to be replaced with new tissue about once every month. A lifetime of flaking removes about 105 pounds of old skin and sees a thousand new outer layers.

Each square inch of skin is made up of 19 million cells, 625 glands for sweat, and 90 glands for oil. Nineteen feet of intricately woven blood vessels serve the square-inch area, along with 19,000 nerve cells. And a stand of some sixty-five hairs is populated by tens of millions of microscopic mites and bacteria. But the outer layer (epidermis) has no life of its own. This external shroud is dead, waiting to be flaked away and replaced by living skin below. □

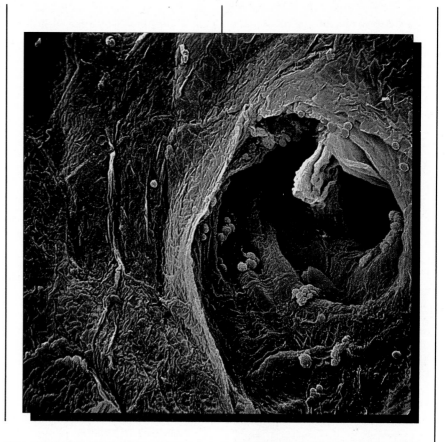

Heavy Breather

The lungs are twin sacks of tissue suspended behind the heart, lined with some 375 million tiny air cells called alveoli *(right)*. Acting as minuscule valves at the end of air tubes, or bronchi, the alveoli effect the vital gas exchange that purges exhausted blood of its carbon dioxide and invigorates it with the fresh oxygen being inhaled.

The exchange itself occurs automatically, propelled by a simple physical law: In gases, high pressure flows toward low. Because there is more oxygen in the inhaled air than in the depleted blood, oxygen flows through the alveoli to the red cells. But since the pint of natural air in the lungs contains only traces of carbon dioxide, the incoming blood readily yields that waste gas to the lung, which ejects it in an exhalation.

Although they occupy only two tenths of a cubic foot, the lungs have an interior surface area of some 861 square feet. In a room of that size, the skin—the body's largest organ by weight—would be a four-by-five-foot throw rug. □

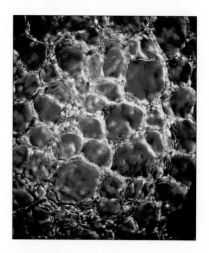

Rude Cauldron

Food taken in by the body must be rendered into a water-soluble form that can travel in the blood. This task of turning chewed-up morsels into chyme, a slurry of liquefied solids, falls to the stomach, a versatile reservoir shaped, appropriately enough, like a horn of plenty. A soft sack formed by a widening at the nether end of the esophagus, sheathed in three layers of muscles, it seems to want only to be filled and emptied. But the stomach is no mere flaccid bag; it is a sturdy crucible.

The stomach's velvety lining contains millions of cells excreting gastric juices—a mixture of water, inorganic salts, mucus matter, and proteins called enzymes—which drive, or catalyze, chemical reactions. The most potent ingredient, however, is also one of the body's oddest fabrications: hydrochloric acid, the same powerful solvent used in industry to dissolve metal oxides—the chemical that makes the stomach a veritable cauldron of digestion.

The presence of such a strong acid in the body calls for a complex chemical balancing act to prevent damage to the stomach. Only the presence of a second type of cell, which smears the lining with a protective mucus layer, keeps the hungry organ from swiftly consuming itself. Despite this defense, however, the stomach lining loses and must replace half a million cells a minute, and the entire lining must be renewed every three days.

A surge in acid production can sear the inner lining horribly, burning the painful lesion called an ulcer in the vulnerable stomach wall: Although the precise mechanism of such imbalances is not known, it appears that a rogue stomach, like a hearty appetite, may begin in the brain. There, just as the mere memory of some delicacy can set the stomach churning in anticipation, emotional stress can spring open the floodgates of hydrochloric acid. □

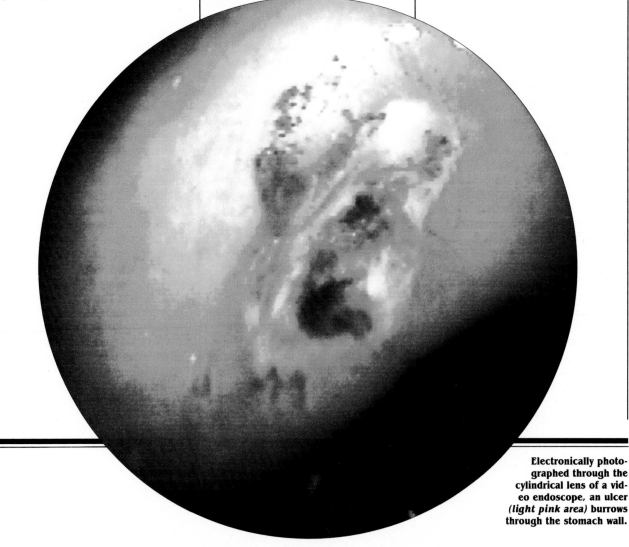

Electronically photographed through the cylindrical lens of a video endoscope, an ulcer (light pink area) burrows through the stomach wall.

Each minute, 300 million of the body's cells die. If they were not replaced, all would be dead in about 230 days. Shortest-lived are the cells of the intestinal tract, lasting only a day or two.

A heart cell begins to beat about six weeks after conception. When a second cell joins it, the two synchronize their beating and become a rough version of the finished organ.

Hair covers virtually the entire body but is thickest and grows most rapidly on the head, which wears a forest of perhaps 100,000 individual hairs. Each lives four to six years, then falls out, to be replaced by a new one, which grows about half an inch a month. Hair also seems to know when to stop growing. During World War II, a skin graft from a soldier's scalp produced a stand of long hair on his repaired thumb. Years afterward, though, the thumb hair vanished, the graft marching in step with the balding veteran's scalp.

Only seventeen muscles work to make a smile, while a frown requires forty-three.

Water makes up 20 to 80 percent of most bodily tissues and as much as 85 percent of brain tissue. If all the water were drained from a 160-pound man, his dehydrated remains would weigh only 64 pounds.

Inner Strife

In a microscopic world beneath the body's tranquil surface, an endless, savage war unfolds, as an internal army and invading enemies struggle to control vast resources of energy, oxygen, and organic matter. Although scientists have known about this war for many years, only in the past decade or two have they come to understand the full scope of the defensive forces and strategies that respond to the body's call to arms.

The battlefield of bloodstream and cell tissue is constantly patrolled by white blood corpuscles, the first-line defenders against bacteria, protozoa, and viruses, and cancerous compatriot cells. Although these warriors are closely connected through chemical communications links, they answer to no central authority. Their clashes occur in a kind of biological police state, where one in every hundred cells is involved in defense.

Every cell in the body carries on its surface three-dimensional identification cards fashioned from specialized proteins, the organic building blocks of life. These badges of identity, called antigens, are the cell's safe-conduct in the body. Without them or with antigens altered by damage or disease, the cell's fate is swiftly sealed.

Like river patrols in a vast delta, white blood cells called phagocytes—produced in bone marrow—roam the body, checking antigens. En route, they sweep up debris and such lethal inorganic substances as asbestos fibers, on which a grazing white corpuscle has impaled itself in the microphotograph on page 20. But *phagocyte* means "cell-eater," and that is how they do their most important defensive work. When they come upon an antigen that is not of the body, they quickly extend a false foot, or pseudopod, and engulf the intruding agent of disease *(see page 21)*. They display the digested alien's antigen on their bodies to alert their cell-eating comrades in arms. They also alert another kind of white blood cell that is trained for special missions.

This cell is known as a T-lymphocyte, named for having passed through the thymus, a small, butterfly-shaped organ at the base of the neck. Many millions of these T cells enter the thymus, but only five percent of them emerge into the body's service. In ways not fully understood, they are selected for their ability to recognize antigens. Collectively, the T cells can interpret millions of antigens, including the friendly ones of the body's other cells and a host of alien identities.

The phagocytes' alarms bring a first wave of helper T cells to im-

In the earliest developmental stages of the human embryo, cells are capable of becoming anything at all. A budding neuron may thus be put to work temporarily as liver or skin, and one destined to labor in the blood may flourish for a while in the cerebral cortex. But specialization comes quickly. A few weeks into pregnancy, a spurt of embryonic chemicals irreversibly commits the cells to what will be their lifelong calling.

The body grows about a third of an inch every night but shrinks to its original size the next day. Gravity makes the difference. In bed, the cartilage disks of the spine are relieved of gravity's downward pull and expand, adding to body length. A prolonged absence of gravity, such as that experienced in space, can stretch an astronaut an inch or two.

The body has about 100 trillion cells—as many as there would be people on 20,000 Earths.

At birth, human babies have some 350 bones, more than one and a half times the number they will have as adults. Many of the small bones in the infant skeleton eventually fuse to form larger ones, 206 in all.

After birth, neurons cease dividing, although the brain's mass grows until about age seven. Muscle cells divide only when injury requires them to replenish themselves.

mediate readiness. They percolate out into the bloodstream, searching for intruders wearing the antigenic colors of an outsider. When the T-lymphocytes make contact, some move in to help the embattled phagocytes, while others carry a chemical call for reinforcements to the spleen and lymph nodes, which activate a second type of T cell: killers.

Killer T cells multiply ferociously and move to the attack. But instead of striking directly at alien intruders, they chemically inflict a lethal rip in the fabric of the cell occupied by the enemy, killing both friendly cell and invader.

Not all killer T cells receive antigen training in the thymus gland. A third type, called natural killers, circulates endlessly through the bloodstream, quick-strike units capable of targeting not one but many alien antigens. These T cells attack an infected cell immediate-

ly, without waiting for the call of helpers, providing rapid defense against an explosively multiplying enemy—the viruses in a malignant tumor, for example.

Meanwhile, the helpers have also told the lymph nodes to produce a second kind of special infantry, the B cells. These require a day or two to reach the front, for they must fashion weapons called antibodies—Y-shaped protein molecules tailored to fit the specific antigen carried by the invader. Although the B cells attack the intruding bacterium or virus directly, they are not straightforward killers. Instead, they mark the intruders with antibodies, which activate yet another specialized defense force, a unique group of blood proteins called complements. Summoned by the antibody, nine types of complements surround the enemy, and, when all nine are in place, destroy it.

In the end, when the body has won another skirmish with invaders, a fourth type of T cell—a suppressor cell—announces that the war is over and the mobilization should cease. The lymph nodes and spleen shut down their killer cell production, and most of the soldiers in the field die off—to be consumed, with the other litter of battle, by foraging phagocytes.

On both sides of the conflict, the cost is astronomically high. In an engagement of this kind, millions of the body's defenders and cells are killed destroying millions of the enemy. The body does not readily forget the carnage. Many of the demobilized T cells continue to patrol, becoming memory cells in which the antigen of the attacker may be remembered for a lifetime after the initial confrontation. As long as the T cells recall the intruder's alien flag, the body will be immune to that enemy. □

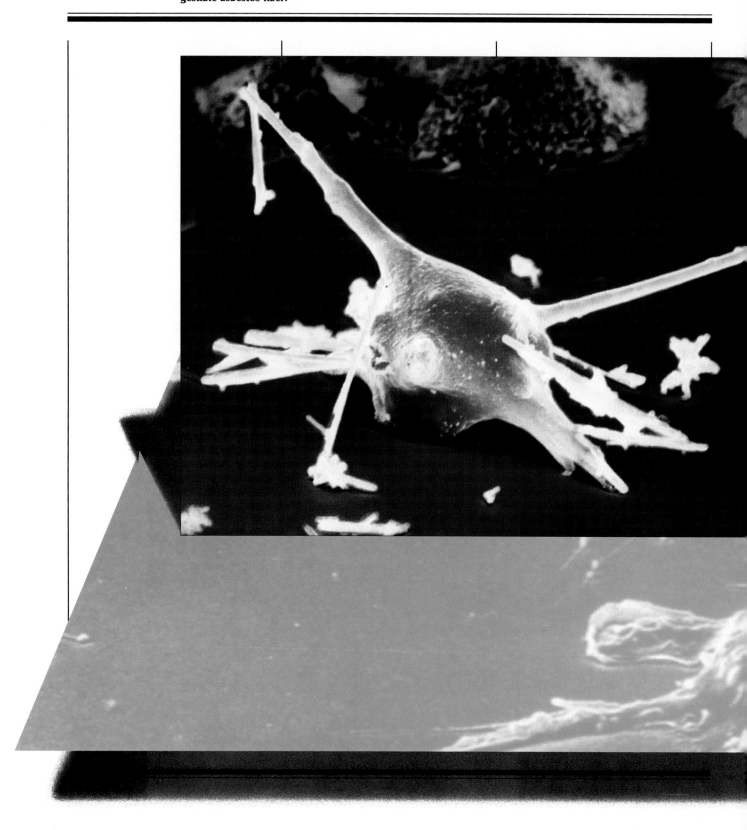

Shown against a backdrop of white corpuscles *(silvery pink)*, immune-system defenders fight intruders. Below, a macrophage, a type of phagocyte, has engulfed—and may be killed by—an indigestible asbestos fiber.

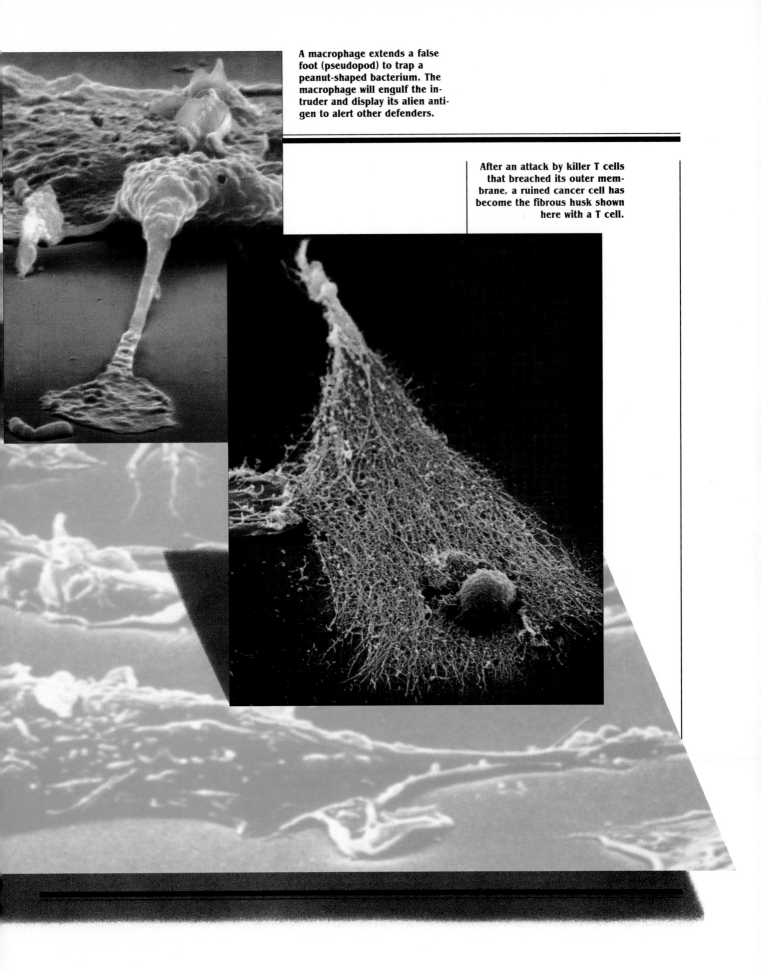

A macrophage extends a false foot (pseudopod) to trap a peanut-shaped bacterium. The macrophage will engulf the intruder and display its alien antigen to alert other defenders.

After an attack by killer T cells that breached its outer membrane, a ruined cancer cell has become the fibrous husk shown here with a T cell.

Terrible Revolution

The show-no-mercy ethic of the body's immune system works well against invaders from the outside world. But it can have a lethal side effect when the same single-minded defenders mistake some of the body's own tissue for the enemy and begin an assault.

The fragile barrier between defense and self-destruction is called immune tolerance, the ability of immune-system cells to tolerate the antigens of other cells native to the body. When the warrior cells lose that tolerance, they sense that they are surrounded by enemies and mount what is called an autoimmune response: the body combating itself. The war is prosecuted as vigorously as it would be against any external invader. Phagocytes devour cells that they perceive to be infected; T cells, B cells, and antibodies target their compatriots.

Every tissue in the body is potentially vulnerable to autoimmune diseases. Systemic lupus erythematosus, for example, attacks cells of the kidneys, lungs, skin, brain, and bone joints. Multiple sclerosis targets the central nervous system; myasthenia gravis, the skeletal muscles; Hashimoto's disease, the thyroid gland. Goodpasture's syndrome is an autoimmune assault upon the kidneys. Rheumatoid arthritis moves against the bone and tissue of joints, causing them to swell painfully and eventually become rigid.

Treatment for such disorders is often as dangerous as the disease, for it requires suppression of the entire immune system. And in an undefended body, even the common cold can be fatal. □

Clever Enemies

Bacteria, protozoa, and viruses, the chief external enemies of the body, have all evolved strategies that are calculated to get around the immune system's formidable defenses. Bacteria, which are single-celled organisms, are able to thrive in the body and produce the deadly toxins responsible for most epidemic human diseases. Bubonic plague, typhoid, dysentery, tuberculosis, leprosy, syphilis, diphtheria, and cholera are all caused by bacterial poisons.

These microorganisms pose a dilemma for the immune system, which must destroy bacterial invaders without harming friendly bacteria that live in the body and play vital roles in digestion and other processes. To discriminate between friend and foe, white blood cells have evolved the ability to interpret antigens by the million. But the world contains tens of millions of bacterial antigens.

Inevitably, some are not recognized as enemies and slip through the immune-system patrols, carrying infection with them.

Protozoa are the greatest threats to health in the less-developed Third World, where they impart malaria, amoebic dysentery, and African sleeping sickness, among other diseases. Stealth is their stock in trade. The malaria agent, for example, reproduces in the liver, then sends its offspring out into the bloodstream at intervals to do their debilitating work. The protozoa behind sleeping sickness and the heart-damaging illness called Chagas' disease change their antigens periodically. Each time their surface antigen is identified as an enemy, the protozoa alter it, permitting them to keep one step ahead of the immune system. These patient parasites also wait for the immune system to weaken, through stress, malnutrition, or such disorders as acquired immune deficiency syndrome (AIDS). Then, even such curable protozoa-borne illnesses as pneumonia become as lethal as any plague.

There is no more potent enemy than the virus. Tiny bodies consisting of only a protein wrapper around a set of genetic instructions, viruses have no real life of their own. They are inert and cannot reproduce by themselves. But,

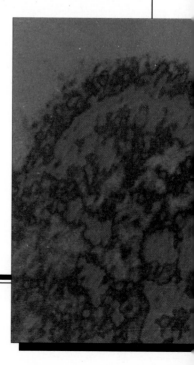

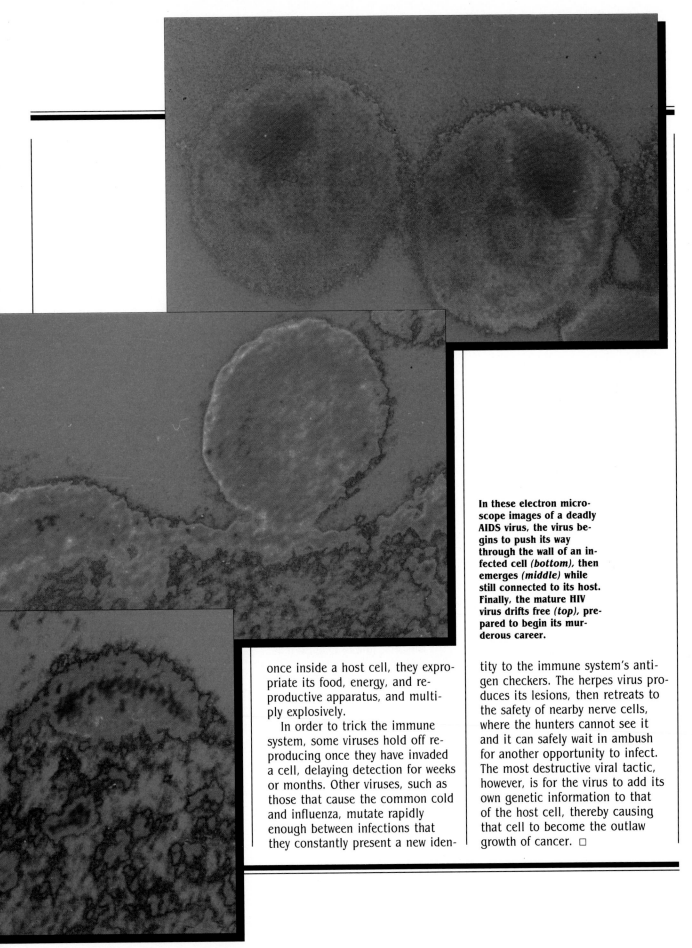

In these electron microscope images of a deadly AIDS virus, the virus begins to push its way through the wall of an infected cell *(bottom)*, then emerges *(middle)* while still connected to its host. Finally, the mature HIV virus drifts free *(top)*, prepared to begin its murderous career.

once inside a host cell, they expropriate its food, energy, and reproductive apparatus, and multiply explosively.

In order to trick the immune system, some viruses hold off reproducing once they have invaded a cell, delaying detection for weeks or months. Other viruses, such as those that cause the common cold and influenza, mutate rapidly enough between infections that they constantly present a new iden-

tity to the immune system's antigen checkers. The herpes virus produces its lesions, then retreats to the safety of nearby nerve cells, where the hunters cannot see it and it can safely wait in ambush for another opportunity to infect. The most destructive viral tactic, however, is for the virus to add its own genetic information to that of the host cell, thereby causing that cell to become the outlaw growth of cancer. □

Tiny Strangers

Although alien invasion of the human body most often poses a threat, sometimes it is the key to perpetration of the species. Human semen, as much an outsider as any bacterium, enters a body to fertilize, leaving behind an object that grows rapidly in the womb. Since the immune system would quickly destroy sperm and fetus if it found them, the body has evolved ways to keep its army in the dark.

Sperm arrive coated with a protective substance that conceals their antigens from the female's phagocyte patrols. The coating also seems to offer protection against the activity of female T and B cells. The mother's body collaborates in this, applying a second coat that further masks the sperm's intolerable antigens.

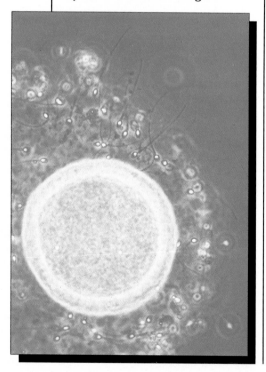

The sperm cell's mission is a short one, its exposure brief. But the fertilized egg faces a more persistent danger. Each cell of the fetus wears the insignia of the mother and father, making it legal game for the mother's immune system, which travels in the same blood with which the fetus must be nurtured in the womb.

Connected to the mother through an umbilical cord, the baby is shielded by a sophisticated filter in the placenta, the branching of the umbilicus where it attaches to the wall of the womb. Although the bloodstreams of mother and child converge here, they do not quite meet. Oxygen and nutrients from the parent's blood are filtered through an inner lining of the placenta called the trophoblast, which removes or chemically disguises fetal antigens that the mother's immune system would otherwise attack. The trophoblast has other stealth tactics as well, launching decoy antigens into the farther reaches of the womb to draw the attention of killer cells. At the same time, it admits antibodies against certain bacterial enemies, passing the immunity of the mother on to the unborn child.

If this first line of defense fails, the fetus is far from powerless. It can produce substances that encourage the mother to generate suppressor T cells, which turn off her immune response.

Ironically, the defensive strategies of human beings at the beginning of life resemble those employed by a less welcome intruder. It appears that some cancer cells, like human fetuses, somehow persuade the host's immune system to call a false truce. □

Like microscopic star fighters, two long-tailed sperm dive toward the planetlike surface of a female ovum.

THE LIMITS OF ENDURANCE

During the 1968 Olympics, American Bob Beamon sprang twenty-nine feet two inches in the long jump. Many who saw his performance distrusted their own eyes; no one had ever jumped even twenty-eight feet before, and it seemed unthinkable that previous records could be so soundly shattered.

Beamon's mark, which still stands, was also a demonstration of a universal human reflex to ignore supposed physical limitations. People have lived after lying, presumably drowned, for half an hour at the bottom of a frozen lake. Some have survived seemingly lethal injuries to brain and body, or endured long periods without food, water, or sleep. Others have fallen three miles out of the sky and lived to describe the experience.

The very stuff of the body helps it to beat the odds. Strong, lightweight materials may bend without breaking, or possess great hardness or tensile strength. A skillfully designed scaffolding of durable bone, for example, accepts hundreds of pounds of pressure and can withstand temporary forces that effectively increase the body's weight a hundredfold.

Internal functions are also managed to preserve life. In extremity, the flow of blood and nutrients adjusts to protect the vital heart and brain. Submerged, the body shuts down its breathing reflex. Faced with a harsh environment, it evolves adaptations to draw oxygen from thin air, insulate itself from the cold, or reduce its temperature.

But the ability to stretch the limits of human endurance is more than engineering or evolution. It springs from a powerful human impulse—the will to continue and prevail.

Fearsome Falls

One afternoon in 1977, a twenty-two-year-old Floridian and novice skydiver named Mark Mongillo *(right)* was performing his twelfth jump when he became entangled in both the main and the reserve chutes. He was at 2,500 feet and hurtling toward earth. "My exact words when the second chute didn't open," he recalled, "were 'Shoot, I'm dead,' and I looked up and said 'God help me.' "

There was nothing to break Mongillo's fall. He hit solid ground, bounced twice, and came to rest in an irrigation ditch. He knows this because he never lost consciousness. He needed eight hours of surgery for a badly broken leg and numerous internal injuries, but there were few psychological scars to heal. "It was a calm feeling," he said of the fall he took. "Something came over me like everything was going to be all right, and I couldn't picture myself dead, you know? It didn't bother me."

For Mongillo, the nightmarish fear of falling from great heights had become sickening reality. But in living through his plunge, he joined a select group of survivors who have escaped death after falls even more spectacular than his own. In January 1942, for example, during the German invasion of Russia, a Soviet air force lieutenant named I. M. Chisov fell nearly 22,000 feet—without benefit of parachute—from his damaged Ilyushin-4 bomber. His landing spot was the snow-covered slope of a ravine, which broke his fall and allowed him to slide to the bottom with a fractured pelvis and severe spinal damage.

Almost exactly a year later, an American airman, Alan E. Magee, survived a similar descent from the same 22,000-foot altitude when his B-17 was hit during a bombing run over France. He smashed through the glass roof of the St.-Nazaire railroad station. Dumbfounded German occupation troops gathered him up and took him to the hospital, where he stayed for three months.

Magee had no recollection of his fall. The last thing he could remember was being sucked out of the blazing plane. But British tail gunner Nicholas Alkemade (above) had no trouble recalling the midnight when his bomber exploded into flames over Germany's Ruhr

Valley in 1944. A pursuing German fighter had scored a direct hit, and Sergeant Alkemade's parachute, stored on its rack at the back of the fuselage, was turned into a heap of cinders. So, 18,000 feet up, he jumped.

"I felt a strange peace away from the shriveling heat," he said later. "If this was dying, it was nothing to be afraid of." A stand of fir trees blocked the impact of his fall, and he wound up in a foot and a half of snow with only minor injuries. Incredulous at his story, Alkemade's German captors planned to execute him as a spy until he showed them that the snap hooks on his parachute harness had gone unused.

The serene, even euphoric, feeling described by Alkemade and

Mongillo is apparently how most people react to a great fall. In 1892, a geologist at Switzerland's University of Zurich, Albert von St. Gallen Heim, published a study that was based on hundreds of interviews with mountaineers, roofers, and others who had fallen appreciable distances. Almost 95 percent of the time, he wrote, "no grief was felt, nor was there paralyzing fright of the sort that can happen in instances of lesser danger. There was no anxiety, no trace of despair, no pain, but rather calm seriousness, profound acceptance, and a dominant mental quickness and sense of surety."

Moreover, he found, "the person falling often heard beautiful music and fell in a superbly blue heaven containing roseate cloudlets." □

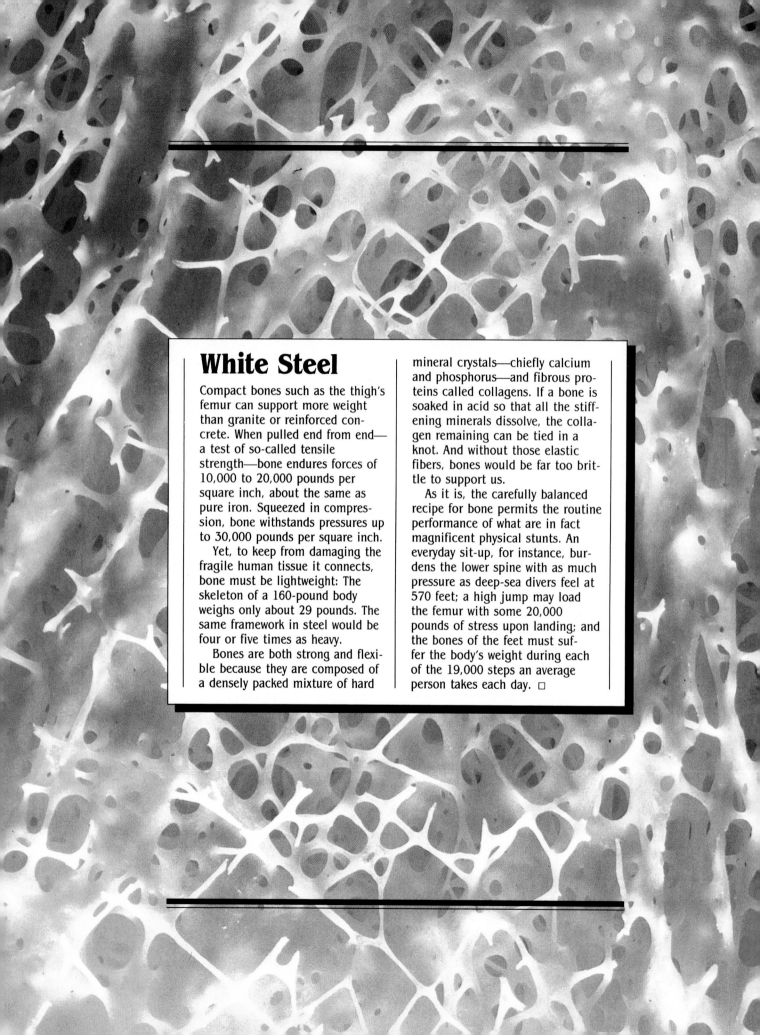

White Steel

Compact bones such as the thigh's femur can support more weight than granite or reinforced concrete. When pulled end from end—a test of so-called tensile strength—bone endures forces of 10,000 to 20,000 pounds per square inch, about the same as pure iron. Squeezed in compression, bone withstands pressures up to 30,000 pounds per square inch.

Yet, to keep from damaging the fragile human tissue it connects, bone must be lightweight: The skeleton of a 160-pound body weighs only about 29 pounds. The same framework in steel would be four or five times as heavy.

Bones are both strong and flexible because they are composed of a densely packed mixture of hard mineral crystals—chiefly calcium and phosphorus—and fibrous proteins called collagens. If a bone is soaked in acid so that all the stiffening minerals dissolve, the collagen remaining can be tied in a knot. And without those elastic fibers, bones would be far too brittle to support us.

As it is, the carefully balanced recipe for bone permits the routine performance of what are in fact magnificent physical stunts. An everyday sit-up, for instance, burdens the lower spine with as much pressure as deep-sea divers feel at 570 feet; a high jump may load the femur with some 20,000 pounds of stress upon landing; and the bones of the feet must suffer the body's weight during each of the 19,000 steps an average person takes each day. □

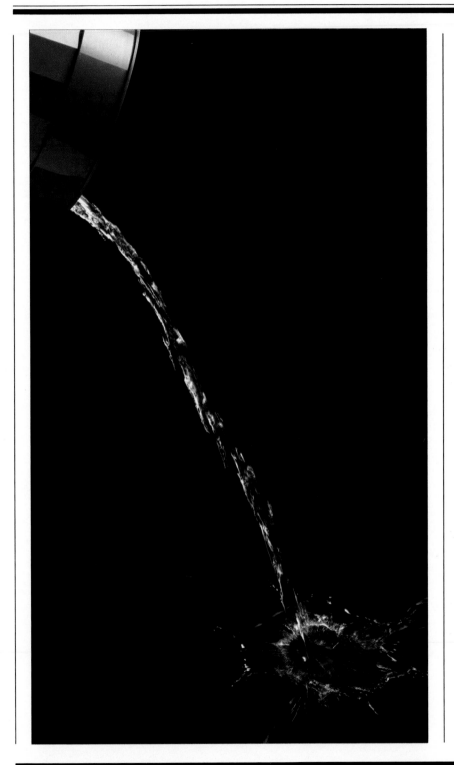

Waterlogged

Nothing is more vital to human life than water. It constitutes 61 percent of the body, or about fifty quarts in an average man. About two and a half quarts are lost each day through exhalation, perspiration, and excretion. If this loss goes unreplenished and the body loses seven to ten quarts of water, death is inevitable. Although people have reportedly survived for months without food, no one has been able to live more than eleven days without water.

But this vital elixir is also a powerful solvent that can upset the body's mineral balance. In rare cases, excessive water intake can dilute the mineral concentrations in the fluid outside body cells; this fluid enters the cells, causing them to swell. Usually, the result is an uncomfortable bloating. When brain cells become thus engorged, however, the result may be an excruciating headache, convulsions, or even coma. □

Bent Nails

Fingernails, which may seem brittle and easily broken, are in fact remarkably tough. A form of bone, they have been described by physiologists as reptilian scales in the process of evolving into hair. Nails are 96 percent protein, arranged in lengthwise, twisting strands—a structure forty times more resistant to fracture than stone. As all nail-biters know, nails bend considerably before breaking. □

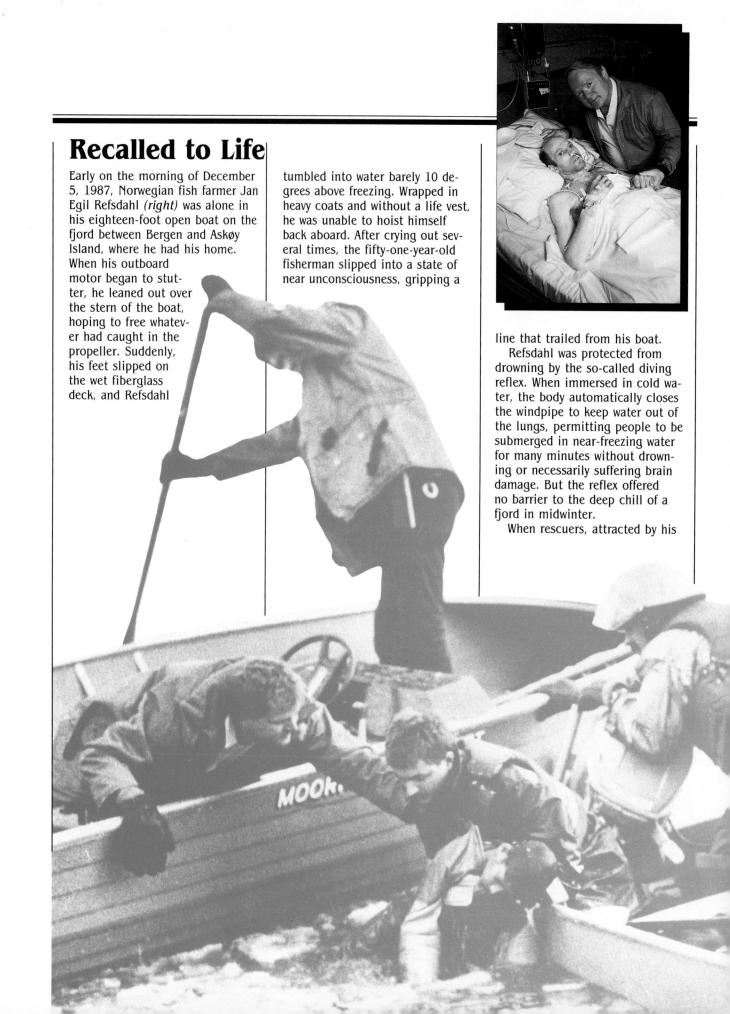

Recalled to Life

Early on the morning of December 5, 1987, Norwegian fish farmer Jan Egil Refsdahl *(right)* was alone in his eighteen-foot open boat on the fjord between Bergen and Askøy Island, where he had his home. When his outboard motor began to stutter, he leaned out over the stern of the boat, hoping to free whatever had caught in the propeller. Suddenly, his feet slipped on the wet fiberglass deck, and Refsdahl

tumbled into water barely 10 degrees above freezing. Wrapped in heavy coats and without a life vest, he was unable to hoist himself back aboard. After crying out several times, the fifty-one-year-old fisherman slipped into a state of near unconsciousness, gripping a

line that trailed from his boat.

Refsdahl was protected from drowning by the so-called diving reflex. When immersed in cold water, the body automatically closes the windpipe to keep water out of the lungs, permitting people to be submerged in near-freezing water for many minutes without drowning or necessarily suffering brain damage. But the reflex offered no barrier to the deep chill of a fjord in midwinter.

When rescuers, attracted by his

cries, arrived, Refsdahl had drifted below the surface, apparently lifeless. Hauling him out of the wintry water, they detected neither pulse nor breathing. An electrocardiogram taken in a dockside ambulance indicated no heart activity whatsoever. At nearby Haukeland Hospital, doctors could find no signs of life. Refsdahl's eyes did not respond to light, and his body temperature was a meager 75 degrees Fahrenheit, more than 23 degrees below normal. Connected to a heart-lung machine, the outwardly dead man's respiration and circulation were restarted—and his heart began to beat after four hours of silence.

Jan Egil Refsdahl is one of a small, elite group who should have been killed by winter's cold but survived. Although no one has surpassed the Norwegian's record for interruption of heart activity, three others are known to have recovered after their bodies chilled far lower than Refsdahl's—the lowest to 60.8 degrees Fahrenheit.

Two-year-old Vickie Mary Davis was found unconscious in an unheated house in Marshalltown, Iowa, in 1956, and showed no vital signs when discovered. And in

1985, Milwaukee toddler Michael Troche *(below)* slipped out of his family's home into a snowy night that left him frozen stiff, his heartbeat silenced. Both children were revived, possibly because their smaller brains and less-developed bodies required relatively little oxygen to function.

But youth does not explain the case of Dorothy Mae Stevens *(right)*, who passed out in a Chicago alley one night in February 1951. There she slept all night, while the mercury dropped to 11

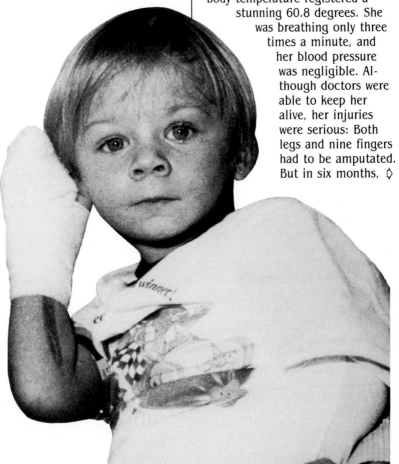

below zero. When she was discovered and brought to a hospital, her body temperature registered a stunning 60.8 degrees. She was breathing only three times a minute, and her blood pressure was negligible. Although doctors were able to keep her alive, her injuries were serious: Both legs and nine fingers had to be amputated. But in six months, ◊

Rescue workers extract eleven-year-old Alvaro Garza, Jr., from the ice-crusted Red River near Fargo, North Dakota, in early December 1987. Although clinically dead after forty-five minutes' immersion in the icy water, young Garza recovered in time for Christmas.

Dorothy Mae Stevens was able to leave the hospital, and she lived for twenty-three more years.

Such extraordinary cases attest to the fact that the body is not just a passive victim of cold. With the first feeling of chill, it has already activated a strategy for preserving as much life-sustaining warmth as possible. In the most desperate circumstances, the body subjects itself to a kind of triage, sorting its priorities to ensure survival—deciding what to sacrifice and what to save.

In minor cold, the body first shuts off such cooling mechanisms as perspiration and then begins to shiver. This involuntary flexing of small muscles temporarily increases bodily heat output by as much as 500 percent.

If this is not enough to halt the loss of heat, the body's temperature begins to drop. When it falls below 90 degrees, all sensation of cold vanishes and shivering ceases. Now the chilled body struggles to hoard warm blood deep within to protect vital organs. Surface blood vessels retreat into lower tissue; capillaries in the skin narrow, restricting blood flow. As the skin freezes, blood vessels near the surface fracture, constricting the supply of warm blood to the hands and feet. Fingers and toes may turn black and die from lack of oxygen-bearing blood.

Finally, in the ultimate tactic of survival, the body slows its internal energy production, intent on keeping the brain alive even at the expense of the rest of the body. Few people survive this stage. □

Charmed Brains

Phineas P. Gage had no reason to expect anything special on the morning of September 13, 1848. As a construction foreman for the Rutland and Burlington Railroad, Gage planned to spend the day preparing some territory for the laying of new track. Instead, by that evening, he had gained local fame as the charmed survivor of one of the most bizarre and brutal accidents in medical annals.

The twenty-five-year-old railroad man was packing a load of explosives into the earth near Cavendish, Vermont, when the charge detonated prematurely. Propelled by the explosion, the iron tamping rod he had been using, three feet seven inches long and thirteen pounds in weight, flew backward out of his hands and blasted into his left cheek just above the jaw line. The rod drove straight through Gage's skull and brain, emerging at the crown of his head.

His fearful coworkers took him by oxcart to a hotel a mile away, where two doctors met them. While the physicians examined Gage and cleaned his ghastly wound, he never lost consciousness. Over the subsequent weeks, he hemorrhaged severely, suffered intermittent bouts of delirium, and eventually lost the sight in his left eye. But Gage survived for thirteen years more, exceeding the expectations of all who treated him. The case was a sensation in medical circles of the day, meriting write-ups in the *American Journal of Medical Science* and the *British Medical Journal*. And Gage himself earned

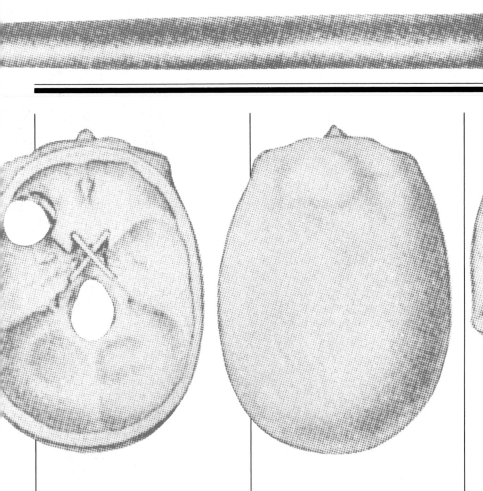

a literal place of honor at Harvard Medical School: His skull and the metal rod that split it *(above)* are on display there at the Warren Anatomical Museum.

Cases such as Phineas Gage's are baffling because the human body is so mortally vulnerable to head injuries, especially those that damage the brain. Still, a charmed few suffer direct, shattering blows to the brain without being killed or even incapacitated—usually because they have not experienced the massive bleeding and swelling that pose such danger in cases of brain injury.

David Wright of Toronto is a case in point. In 1983, the fifty-four-year-old building contractor was renovating an apartment when he tumbled from a ladder while

clutching a power drill. As Wright hit the floor, the drill bit bored into his forehead. He realized at once what had happened and had the presence of mind to know that above all else, he must not move the drill, lest it rupture a vital brain component.

Ever so gingerly, Wright made his way across the apartment so he could stand before a mirror, where, calmly using his own horrific reflection to guide him, he switched the drill into reverse and withdrew it. Then he walked upstairs to his apartment and asked his wife to summon an ambulance. Surgeons who explored his wound found that the drill had penetrated two inches into Wright's head, far enough, certainly, to damage brain cells. But the patient, who—as one

might expect—bore a disfiguring scar from the surgery, managed to recover in full.

The annals of medicine are full of people whose brains were impaled by javelins, crossbow bolts, knife blades, and even umbrella ribs. In 1879, for example, milling machinery drove a heavy bolt four inches into the skull of a woman factory worker, who lost brain tissue but lived another forty-two years without discomfort. When Boston's John Thompson ran his car into a tree in 1981, a forty-pound steel crowbar in the backseat hurtled through his head just above the brainstem to protrude over his left eye. He suffered some paralysis and impaired speech but was up and about three months after the accident. □

Cases of Endurance Unlimited

Human ability to withstand extreme injuries and to endure excruciating pain sometimes seems limitless. Now and then, the physical and emotional demands of accidental injury appear to be unbearable—yet, they are borne.

In 1983, when eighteen-year-old Kimberly Lotti *(below)* accidentally swerved her pickup truck into a chain-link fence in Quincy, Massachusetts, an aluminum post two inches in diameter speared through the windshield and passed through her chest. She remained conscious as rescuers sawed off both ends of the stake several inches from her body. The rest was removed at a hospital. In a 1985 collision with a car in Derbyshire, England, motorcyclist Richard Topps was hurled through the air and impaled from chest to hip on a wooden

post—where he dangled helplessly for more than an hour before being found by his brother.

Both Lotti and Topps were lucky; their internal organs were undamaged. A Colorado man who fell from a window in 1987 had a worse time of it. A metal fence post ripped through his abdomen and entered his heart. Luckily, rescuers realized they had better not remove the spike: Like a finger in a dike, it kept blood from gushing out. His heart never missed a beat,

and surgeons were able to repair the damaged organ.

Even more astounding was the case of Forthman Murff, a seventy-four-year-old Mississippi lumberjack. In 1984, a falling branch knocked him backward into a ditch and onto his own whirring chainsaw. The sawteeth ripped into his throat but somehow missed his spine and carotid artery. Murff kept his composure. Bending over every few minutes to drain blood from his lungs in order to breathe, he drove several miles to the home of a friend, who took him to a hospital another hour away. Doctors who stitched Murff back together had never seen the like of it: The old lumberjack's neck had been almost completely severed. □

Traumas

When a patient is hurried into a hospital emergency room with a bullet or knife wound, chances of survival depend on a host of circumstances, foremost among them the subject's age and physical condition and the skills of the medical team. But a quick look can provide some idea of the victim's odds, based on statistics compiled by medics at Denver General Hospital.

As would be expected, head wounds carry the highest risk, but they are not invariably fatal. If a bullet enters only one hemisphere of the brain, the patient has a slight, one to two percent, chance of surviving. The most famous recent instance of such a recovery was that of James Brady, the White House press secretary wounded in an attempted assassination of President Ronald Reagan in 1981. If the bullet crosses the brain and penetrates both hemispheres, however, death is a virtual certainty.

Gunshot and stab wounds to the heart are extremely dangerous but not always lethal. In a Denver study conducted between 1980 and 1988, 30 percent of the heart-wound victims—twenty-one out of seventy—survived.

For bullet and knife wounds in the rest of the body, survival chances show a remarkable improvement. Studying a sample of 203 patients between 1981 and 1984, divided about equally between victims of shootings and victims of stabbings, Denver researchers found that 94 percent survived if they showed vital signs when first seen by paramedics at the scene of the assault.

In contrast, victims of blunt trauma—wounds from objects such as clubs that do not enter the body—have an 85 percent likelihood of recovering after surgery. □

Demonstrating the destructive trauma of gunshot wounds, a high-velocity bullet blasts through an apple in this strobe-lighted photograph.

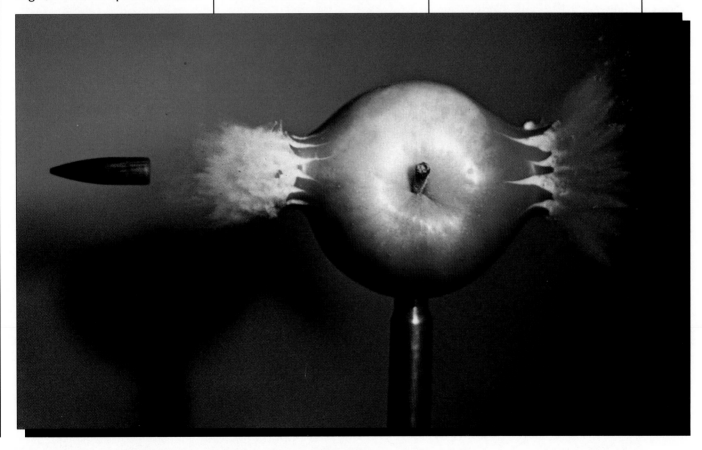

Pushing the Limits

On May 6, 1954, fighting a gusty cross wind and a cinder track still damp from a rain shower, Roger Bannister ran a mile in three minutes fifty-nine and four-tenths seconds (3:59.4). The twenty-five-year-old British medical student had finally broken the mark that runners had talked about and aimed at for decades: the four-minute barrier. Amid the general jubilation, one slightly dissonant note was sounded, by the man who had set the record in the same event twenty-nine years earlier. "What's a world record now?" scoffed Paavo Nurmi of Finland, whose record time was 4:10.4. "It's a printed document at which everybody sets a target and soon it's cracked. Soon even a four-minute mile won't be a rarity."

However ungraciously, Nurmi spoke the truth. By 1986, one track historian calculated that 450 men—no women yet—had duplicated or bettered Bannister's feat. One was Steve Cram of Britain, who in July 1985 became the world's fastest miler at 3:46.32. In that race, Bannister would have finished ninety-seven yards behind. Clearly, athletic records are made to be broken.

Despite the certain knowledge that a new record must fall, the world likes to see one achieved, possibly because in doing so the athlete seems to disregard the physical limitations of being human. Even the most talented competitor must reckon with the restrictions imposed by mechanics and biochemistry.

Distance runners, for example, must stop when the fuel they have stored as carbohydrates runs out.

American Bob Beamon *(left)* makes a twenty-nine-foot, two-and-a-half-inch long jump at the 1968 Olympics in Mexico City—a record that has yet to be broken. Above, the Soviet pole-vaulter Sergei Bubka demonstrates his own record-setting form.

They are also limited by how efficiently they can process oxygen. Unless humans can evolve larger hearts and lungs, researchers say, a mile faster than 3:34 cannot be run. Perhaps they are right; or perhaps their figure is like the six feet eleven inches once seen as the eternal and absolute maximum for a high jumper, which has since been bettered by more than a foot.

For all their arduous training, sprinters are in a sense born, not made. They need a high proportion of special muscle fibers that can only be inherited. These so-called fast-twitch fibers make up more than 70 percent of the leg muscles of Carl Lewis, who walked away from the 1988 Olympics with ◊

one silver and six gold medals in track and field events, and of Florence Griffith Joyner, who smashed the women's 200-meter record that same year. The fibers are fueled by an anaerobic chemical reaction— one that does not use oxygen. Over short distances, Lewis has been clocked at a dizzying speed of twenty-seven miles per hour. If he could maintain that pace, he would run a marathon in sixty minutes. But the anaerobic reaction causes lactic acid to accumulate in muscle tissue, slowing the runner after about 400 yards.

Runners who excel at longer distances have, by contrast, a higher share of slow-twitch, aerobic fibers, which produce no build-up of lactic acid. They then draw on anaerobic fuel for the home-stretch speed needed to win. If a marathoner tires before the finish, it may be because he or she has run too fast too early in the race and carries a lactic acid burden.

Computer simulations indicate that an ideal sprinter who combined fifteen factors at optimum levels could knock a quarter of a second off the current 9.83-second record for the 100-meter dash. But future advances in medicine, nutrition, and techniques of mental discipline should continue to push the limits of athletic possibility.

So may innovations in equipment and performance style. When high jumper Dick Fosbury introduced the backward roll over the bar in 1968, it proved to be a quantum leap for the sport: He soared seven feet four and a quarter inches, two and a half inches higher than the previous world record. Pole-vaulters could only fantasize about reaching sixteen feet until they switched from aluminum to fiberglass poles in the 1960s. Since then, Sergei Bubka of the Soviet Union has broken his own pole-vaulting record eight times, most recently clearing nineteen feet ten and a half inches.

For all that, Roger Bannister, knighted in 1975 by Queen Elizabeth II, still stands in the minds of many as the exemplar of individual achievement. He did not mind relinquishing the record; for quite some time, he sent a commemorative necktie to anyone who beat the four-minute mile. In 1976, having become a successful neurologist, he learned a different kind of lesson about the limits of the body. His car was struck by another at a London intersection, and his ankle was badly fractured. Roger Bannister, the quintessential runner, had to give up his sport. □

MIRROR MULTIPLES

For many primitive peoples, the birth of two or more identical offspring presaged such evil that the hapless babies were often destroyed to placate the gods. But modern society finds the phenomena of identical twins, triplets, quadruplets, and quintuplets mainly fascinating, a seeming prank of nature that renders still more miraculous the miracle of conception.

Part of the interest in multiple births comes from the fact that they are uncommon. Although there are an estimated 62 million sets of twins in the world today, one-third of whom are identical, 99 percent of all births yield single offspring, and only 1 birth in 250 produces an identical pair. Such identical copies arise from a single human ovum, or egg, which divides after fertilization to form two or more genetically indistinguishable embryos. (Fraternal twins, who do not necessarily resemble each other, come from two eggs.)

The profound linkage between like human beings far exceeds anything in the experience of the untwinned. Identical offspring often share a secret language in childhood, behave as composite personalities as they grow up, and forge deep bonds with one another. Some report feeling sympathetic pain when their doubles hurt, while others claim the ability to communicate with their matched siblings telepathically. Identical twins separated at birth and reared apart may display eerie similarities in behavior that even the skeptical acknowledge cannot be attributed to mere coincidence. It is as though each copied person follows a kind of genetic predestination, unable from the moment of conception to mature completely independent of the other. But nothing can quite match the experience of physically conjoined Siamese twins. Formed when a single ovum fails to separate completely, they are often linked by the same limbs or internal organs, and together must endure their destiny.

Britain's Silent Twins

The closeness of identical twins is sometimes envied by those who long for a soulmate with whom to share deepest secrets and most private thoughts. For twins, however, such intimacy can occasionally be psychologically devastating, as it has been for June and Jennifer Gibbons, Britain's famous Silent Twins. Born in 1963 in the British crown colony of Aden, where their father—a native of Barbados—was serving in the Royal Air Force, they made a determined effort as they grew up to live in the world strictly on their own terms. Eventually, their determination would bring the pair to an indefinite sentence in Broadmoor, Britain's notorious maximum security hospital for the criminally insane.

From childhood, June and Jennifer shut themselves away in a world of their own construction, bonded to one another by love—and, strangely, hatred. Although in time they became capable of speaking clearly and articulately, they chose silence when around adults and a clipped, staccato English only they could understand when they wished to communicate with each other. They were what specialists call elective mutes.

The girls' mother had worried when they were slow to develop normal language skills as toddlers but had not been overly concerned. She believed, as did some of their teachers, that her daughters were simply shy. But so strong was their attachment to each other and so complete would their detachment from everyone else become that by their early teens June and Jennifer had retreated to the sanctuary of their bedroom. If they had some-thing to communicate to other family members, they wrote notes. They even ate in their room, accepting food left outside the door by their parents.

Although they did little more than grunt or nod in the company of others, they could be noisy when by themselves. The household could sometimes hear giggles and the sound of furious fighting coming from the twins' room.

When they were thirteen, June and Jennifer were sent for psychotherapy to a live-in treatment center. There they became immobile for hours at a stretch. Their therapist, who remembered later the difficulty of rousing June from bed in the mornings, said that it felt as if he were "lifting out a plank and propping it against a wall."

Nine months of treatment at the center proved to be of no avail. Recalled one of the team of mental-health professionals who had attempted to break through to the girls, "Everyone who came near them felt their power. Some locals even believed they had the evil eye. They split our team and we often found ourselves quarrelling while they looked on with passive, impervious smiles."

When, at a teacher's suggestion, they were separated briefly and sent to different schools, the twins reacted violently. But instead of lashing out at the teacher, they inexplicably screamed and

scratched at each other's face.

As they grew older, the twins craved notoriety and excitement. They gave up some resistance to being with others, especially young men, and began to venture out— even if this meant speaking. In 1981, they took up with three American brothers from a Navy base and learned about glue sniffing and drugs. Their first real social relationship ended shortly after it began, when the boys left Britain and the twins' friendly letter to them came back unopened. Soon after, the girls started stealing petty objects and setting fires in empty buildings, causing close

to $300,000 in damage. They were about to burn down a local technical college when they were apprehended and charged on sixteen counts of theft and arson.

The detective who questioned June and Jennifer about their activities was forced to communicate as their teachers had—by writing his questions, then leaving the twins alone while they responded, in writing. Yet when a search was made of their room, an amazing discovery was made. Although they had eschewed regular speech for much of their lives, June and Jennifer had passionately embraced the written word. Books stood stacked near their bed, including one entitled *The Art of Conversation*. More sur-

prising still, diaries, poems, stories, letters, and completed novels by the girls turned up, including June's novel, *Pepsi-Cola Addict*, which had been published by a vanity press and paid for with money the twins had received in government welfare payments.

Their writings, set down in an almost illegibly tiny scrawl, ran to more than a million words and were filled with stark, vividly articulated insights. For example, June wrote in 1981: "We are both holding each other back. She does not want jealousy, or envy, or fear from me. She wants us to be equal. There is a murderous gleam in her eye. Dear Lord, I am scared of her. She is not normal. She is having a nervous breakdown. Someone is driving her insane. It is me."

Sent to Broadmoor and separated for a lengthy period for the first time, the twins continued to fill their diaries. "I say to myself how can I get rid of my own shadow?" wrote June. "Without my shadow would I die? Without my shadow would I gain life?"

But there was not much of a life without one's shadow, after all. When British reporter Marjorie Wallace, who has written extensively about the twins, first visited them in 1982, she found them wood- ◊

Eight-year-old Jennifer Gibbons *(left)* sends sister June a silent eye signal. Ten years later, not long before their internment in London's maximum-security mental hospital, June *(above, left)* and her twin were photographed in their boyfriends' house in Wales.

enly rigid and deeply depressed. Only Jennifer tried to speak: "Please tell them to let us be together again." On a subsequent visit, some four years later, Wallace found the girls much improved—to the point that they were allowed to spend time with each other, depending on how well they behaved alone and together. And they even tried to explain why they had been silent all those years: "We tried to speak to our parents," said Jennifer, "but it was more comfortable just nodding our heads; words seemed too much. If we were suddenly to talk, it would be too much of a surprise." □

Twinned to Die

When neighbors of Dr. Cyril Marcus complained about a foul odor pervading the tenth-floor hall of their Manhattan high-rise building during the summer heat of July 1975, they may have sensed that something terrible waited behind the double-locked, barricaded door of apartment 10H. But nothing prepared them for what police discovered when they broke in three days later. A fly-ridden compost of rotting garbage, human waste, empty soda and beer cans, and spent barbiturate vials spread through the once-luxurious apartment. In the bedroom, they found the badly decomposed body of Cyril's identical twin brother, Stewart, lying prone on the bed, dressed in shorts, with a single black sock dangling from his left foot. Nearby, on the floor, lay the naked, better-preserved body of Cyril.

Autopsies showed that Stewart had apparently died of barbiturate intoxication, and some experts speculated that Cyril, whose system showed no trace of drugs, might have succumbed to the convulsive withdrawal symptoms of a barbiturate addict denied his dose. Other students of the psychically conjoined lives of the forty-five-year-old Marcus brothers believe Cyril died because his twin had died; the loss of one proved too much for the survivor. In a sense, they arrived at that fatal moment because they were identical twins.

Always together, they had grown up in a middle-class neighborhood in Bayonne, New Jersey. They even shared parental punishments—no matter who had misbehaved, both were spanked. They attended Syracuse University, where they pledged the same fraternity and had the same circle of friends. They even dissected the same cadaver as first-year medical students. In 1954, they were graduated from medical school with honors and began what looked to be distinguished careers.

By the end of the 1960s, Stewart and Cyril Marcus had become successful gynecologists with a lucrative Park Avenue practice and a national reputation for their innovative research into reproduction and infertility. A text on infertility written by the Marcus brothers enjoyed wide circulation.

In 1972, Cyril's health began to deteriorate, perhaps as the result of a mild stroke. Propping himself up with sedatives and other drugs, he began the descent into addiction. Then, as Cyril's behavior became more and more erratic, Stewart took to covering for him by fooling people into thinking that he was his brother.

But Stewart was in trouble too. Increasingly, he was unable, or unwilling, to resist the dementia that had begun to overtake Cyril. It was as if between them they shared only one identity, inexorably forged by their twinship. Faced with his brother's disintegration, Stewart appears to have been helpless to avoid his own. Irresistibly, he trailed Cyril down the spiral of drugs and self-destruction. The first to enter life, Stewart was first to die. Finally alone, Cyril followed his brother a few days later. □

Brilliant medical careers—and tragedy—await identical twins Cyril *(far left)* and Stewart Marcus, shown in these 1954 yearbook photographs from the State University of New York's Upstate Medical Center in Syracuse.

The Two-Headed Boy

Born in 1877 in northern Italy, Giacamo and Giovanni Battista Tocci were survivors from the start. They were identical twins but came into the world after the twinning process had somehow broken down, leaving their bodies partly fused. Their heads, upper torsos, and arms were quite normal, but from the sixth rib down, they were one, sharing the same abdomen, a single reproductive system, and only two legs, on which they could eventually stand but not walk.

Realizing that a "two-headed boy" was something people would pay to see and that the boys were not likely to survive for long after their birth, their mother allowed the twins to be put on exhibit during their infancy. But instead of succumbing to the frailty characteristic of Siamese twins, Giacomo and Giovanni Battista remained strong and healthy. As their fame and fortunes grew, the brothers learned French and German, and traveled widely with their show. In 1892, when they were fifteen, they toured in America for a reported thousand dollars a week.

On the American tour, the Toccis acquired another kind of fame. *Scientific American* called them "probably the most remarkable human twins that ever approached maturity." And Mark Twain used them as models for fictional characters—conjoined Italian brothers in his farcical story, "Those Extraordinary Twins" and identical but separated twins in a better-known work, *The Tragedy of Pudd'nhead Wilson*.

By their early twenties, however, the Toccis had wearied of being always on display, and they retired to the countryside near Venice. There, married to two sisters, they spent their lives concealed in a walled villa, where they died in 1940, aged sixty-three. □

Billed as a two-headed boy, Giovanni Battista (left) and Giacomo Tocci were in fact two boys with one lower body; they are shown here in 1892 as they began their triumphant American tour.

A Curious Case of Motherhood

When Rosa and Josepha Blazek's mother realized she had given birth to Siamese twins joined at the spine and pelvis, she refused the girls food for several days, believing they would be better off dead. But the twins refused to die, and they eventually became a celebrated attraction in sideshows throughout Europe and America.

Born in Bohemia in 1878, the Czech pair learned how to play the violin and the xylophone and used their talents to charm their audiences. They could also dance, amazing those who came to see them with graceful movements so neatly synchronized that they could waltz with different partners. The unmarried sisters must have been equally charming offstage, for, on April 17, 1910, Rosa (near left) gave birth to a healthy baby boy. Although both initially professed ignorance of any conception, Rosa finally admitted having an affair. Josepha (far left), who maintained her innocence, called her sister a "wicked debauchee" but could not have been completely oblivious: The twins shared portions of the same reproductive system.

Rosa's lover eventually came forward and proposed marriage, but it could not be. The prospective husband was unable to obtain a marriage license because, according to the conventions of the day, such a union would have constituted bigamy. □

Original Twins

Conjoined brothers born in 1811 near Bangkok, Thailand (then called Siam), came to be known the world over as the Siamese Twins, a term that has since become a synonym for unseparated pairs. Named Chang *(far right)* and Eng, they were linked at the breastbone by a tough three-and-a-half-inch-long ligament. Displayed to millions in Europe and America, the pair eventually married the Yeats sisters of North Carolina, where they settled as prosperous farmers to raise their twenty-two children.

One night in January 1874, Eng awakened with a terrifying sense that something awful had happened to his brother. He could not hear Chang's breathing, and no amount of shaking made him stir. Discovering that Chang had died, Eng exclaimed to one of his sons, "My last hour has come." Within two hours, he also was dead. At the time, some doctors believed that Eng found the prospect of life without his brother so terrifying that he died of fright. The actual cause of death was probably a circulatory failure, since the twins shared blood vessels. □

The Hidden Twin

Many people have described a vague, persistent feeling that somewhere, somehow, they have a twin who vanished. Research suggests that this notion may be not quite as fantastic as it sounds, for, it appears now, many more twins—fraternal as well as identical—are conceived than born. In fact, studies indicate that anywhere from 26 to 86 percent of conceptions begin with twin embryos but wind up in single births. More amazing still, the unborn twin may be incorporated by the other.

Nick Hill learned of his vanished twin when he was twenty-one. The Idaho service-station attendant had suffered from near-crippling headaches for years. He finally underwent exploratory brain surgery. Doctors found no tumor but discovered a small accumulation of embryonic tissue—skin, hair, and bone—which they concluded was the remnant of Hill's unborn twin. Apparently, when both were embryos, the healthy expanding embryonic cell mass engulfed or enveloped the weaker one.

Sonography, a technique that uses ultrasonic sound waves to delineate a fetus in the womb, has provided vivid evidence that many pregnancies actually begin as potential multiple births (above, right). In 1973, Dr. Louis M. Hellman, of the State University of New York Medical School in Brooklyn, reported that of 140 women with at-risk pregnancies, 22 began with twin gestational sacs, a 25 percent higher incidence of twinning than had been expected (14 of these pregnancies terminated in miscar-

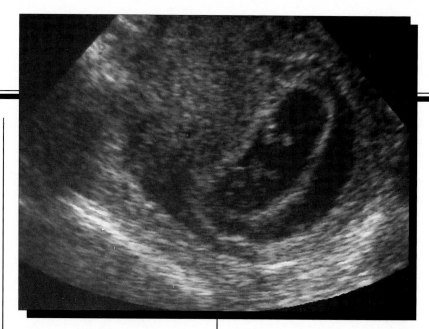

This sonogram reveals twin embryos—the one at left displaying the shape of a human fetus, the other a blurred crescent to its right—during the first trimester of pregnancy. A second sonogram less than six weeks later showed no sign of the less-developed twin.

riages). In 1976, Dr. Salvator Levi, of Brussels' Free University, reported a remarkable statistic based on ultrasonic examinations of almost 7,000 pregnant women. Studied prior to their tenth week of pregnancy, about 71 percent carried twin embryos; but all eventually gave birth to only one child.

According to Levi, the vanished twins in his study were usually gone without a trace by the end of the first trimester (three months) of pregnancy. In most cases, they were resorbed by the mother's body. Some researchers speculate that this may be nature's way of eliminating a flawed fetus without aborting its healthy counterpart.

When the vanished twin is not resorbed but is instead engulfed by the other fetus, portions of it may remain in the living twin's body. As in Nick Hill's case, cysts containing such traces of another infant as teeth, hair, and fingers have been removed from patients. And in one astonishing instance,

an autopsy performed on an old man even revealed a fully developed, six-pound fetus that he had been carrying around inside himself all his life.

Occasionally, the vestigial sibling makes its presence known indirectly. In England, Patricia McDonnell learned when she became pregnant that she had not one blood type, but two, 7 percent type A and 93 percent type O. The type A blood was hers. But most of the blood flowing in her body was that of an unborn brother, whose remains had continued to produce blood after he was engulfed by his sister in the womb. Specialists were able to tell the twin's sex from chromosomes in the type O white blood cells. □

Defying odds that have been estimated at 1 in 50 million, thirty-one-year-old Linda Whicher of Southampton, England, gave birth in 1983 to her third successive set of twins. Afterward, she declared that she and her husband "will not be having any more babies."

Three of a Kind

Robert Shafran became increasingly puzzled during the early weeks of his 1980-1981 freshman year at New York's Sullivan County Community College. Although he was new to the campus, teachers and upperclassmen seemed to know him. It turned out that he was being mistaken for someone named Eddy Galland, who had previously attended the school. After a friend of Eddy's brought the two together, the look-alikes compared birthdays and birthmarks and excitedly concluded that they must be identical twin brothers, adopted by different families.

Ten days later, in the wake of a local-newspaper story about their reunion, Eddy's adoptive mother received a call from a David Kellman, who brought startling news. "You're not going to believe this," he said, "but I believe I'm the third." David was the third. It turned out that Eddy, Robert, and David were identical triplets. Chances of an identical triplet birth occurring are 1 in 100,000.

Born on July 12, 1961, the boys were separated shortly thereafter and adopted by three different sets of parents, none of whom knew about the other boys. Despite their being reared separately, the brothers are uncannily alike in personality. "All my life I have felt special and individual," David said after their against-all-odds reunion. "Now I've met two people just like me." So much alike did the brothers find themselves that they went into business together. Today they own and operate Triplets Roumanian Steakhouse in Manhattan. □

A Sad Quartet

When the quadruplets were born in the early 1930s, there was great public interest and fanfare. The odds of identical quads being born are about 1 in 16 million, and only half of those survive. But the four sisters share more than a common birth—they are also afflicted with the mental disorder schizophrenia, an illness that brings hallucinations, delusions, and psychotic behavior to perhaps 1 out of every 100 Americans.

The sisters have been extensively studied for three decades by psychiatrists at Washington's National Institute of Mental Health, who hope to determine the roles of heredity and environment in mental illness. In order to protect the quads' privacy, the researchers have not revealed their identities, referring to them instead as Nora, Iris, Myra, and Hester—the first letters of which spell the acronym for the National Institute of Mental Health—and the surname Genain, derived from the Greek words for "dire birth."

The Genain quads, now in their fifties, were once portrayed as happy, well-adjusted children by enthusiastic newspaper reports in their Midwestern community. In fact, their lives were dominated by difficult parents. The mother was inflexible and manipulative, the father a hard-drinking, egocentric bully who frequently insisted that his daughters undress in front of him. He never let them play with other children and even forbade their participation in church functions. "There are four of them," he said. "They ought to be able to entertain each other."

Such an upbringing undoubtedly contributed to the onset of the quadruplets' schizophrenia, symptoms of which began to appear in high school and persisted after the girls had gone their separate ways. But NIMH researchers note that the quads' paternal grandmother, great-uncle, and uncle all had nervous breakdowns, and that the four women's brain-wave activity is remarkably similar—clues, perhaps, that their illness may indeed be partly genetic. □

The Quints

On a Sunday night in May 1934, twenty-five-year-old Elzire Dionne, only seven months into her seventh pregnancy, unexpectedly went into labor. At about 4:00 the following morning, May 28, a frail baby girl was delivered, barely alive. By 5:30, four more tiny girls had been born, each weaker than the last. The entire quintet weighed only ten pounds. No one expected them to survive, much less become a popular attraction that some at the time would call the Eighth Wonder of the World.

The Dionne quintuplets—Marie, Annette, Emilie, Cécile, and Yvonne—were born in the family's six-room farmhouse in the small Canadian town of Callander, Ontario, about 180 miles north of Toronto. Word of their arrival quickly spread around the world. Until 1934, only two cases of identical quintuplets had been reported, and none of those babies had survived past infancy.

This sudden doubling of the number of his living children jarred the father, Oliva Dionne, who could barely support the family he already had. Since not even their doctor, Allan Roy Dafoe, believed the quints would live long, Dionne agreed to let them be exhibited—under hospital-like conditions—at the 1934 Chicago World's Fair. Reportedly, he was to receive $80,000 for the show, but when the agreement was made public, it caused a strong and immediate backlash against him and

was called off. Portrayed as exploitative parents, he and his wife were forced to grant temporary custody of the girls to the Canadian Red Cross. Later, by an act of Parliament, the quints were made wards of the state and placed under a four-man board of governors that included Dr. Dafoe but excluded their parents.

Dafoe took over the running of the quints' lives. The girls were tended by white-clad nurses inside a specially constructed, fenced-in compound called Dafoe Hospital, where their activities were closely scrutinized and controlled. Except

for a ten-minute car ride, the girls' first five years were spent totally within the confines of the hospital, under an exacting, gruelling routine that included being shown to the public every afternoon through fine-wire mesh screens.

From the start, the Dionne quintuplets were big business. As one of the biggest human interest stories of the Depression era, they received constant attention from the media, appearing regularly in newsreels and feature films. On their first birthday, scores of telegrams poured in from such celebrities as author F. Scott Fitzgerald, who wrote, "Congratulations on your interesting chapter in the great novel called life."

In 1935 alone, the quints reportedly attracted some $50 million to Ontario, revitalizing the destitute economies of Callander and nearby North Bay. By the time the sisters were ten, more than five million people had waited in long lines at "Quintland" to see the famous Dionnes, whose names and faces promoted dozens of products, ranging from General Motors cars to Karo corn syrup. They were making money not only for their caretakers and themselves, however. A circuslike atmosphere prevailed around the hospital (even their father opened a souvenir stand across from the facility), and the tourist dollars kept rolling in.

But behind the fairy tale, all was not well. Separated from their parents until 1943, when their determined father regained custody of them, the quints were isolated from other children and forced to dress alike for public display. Small wonder that they became increasingly shy and dependent on one another. They knew nothing of the outside world and were ill-equipped for a normal existence. And, although the family was now wealthy, Oliva had become a suspicious and embittered man. Because he was afraid that they would be recognized if they ventured outside, he kept the girls closeted at home just as Dr. Dafoe had in the hospital.

Still, the quints were growing up. On their eighteenth birthday in 1952, their father announced that they would finally leave their cloistered home to attend a convent school. Two years later, Emilie suffocated during an epileptic seizure and died in her bed.

In 1955, when the remaining four girls turned twenty-one and gained their independence, a profound bitterness toward their parents and other siblings surfaced. That Christmas, having failed to receive even a card from the girls, Oliva Dionne called a press conference to chastise his daughters publicly. A subsequent public announcement told the world that it had all been a family misunderstanding. But eight years later, the quints authorized a series of articles and coauthored a book detailing how miserable their lives had been at home. "There was so much more money than love in our existence," they wrote.

Cécile, Marie, and Annette went on to marry, though none of the marriages lasted. In 1970, Marie died alone in her Montreal apartment from a blood clot in her brain. The quints' father died of a stroke in 1979 at the age of seventy-six, and their mother died seven years later at seventy-seven.

The three surviving quints, now in their mid-fifties, live close to each other near Montreal. The trusts set up for them in their youths are mostly depleted. Cécile is a supermarket cashier and Annette and Yvonne are both clerks in a library. □

Strangers No More

Psychologist Thomas J. Bouchard, Jr., would first notice the coincidences at the Minneapolis-St. Paul airport, when he met the identical twins who were participating in a study he was undertaking at the University of Minnesota's Center for Twin and Adoption Research. Normally, Bouchard would have expected similar behavior of such look-alikes. But these were twins who had been separated at birth; they had all grown up with different parents and in different environments, and were still strangers to each other. Yet, as he greeted them, Bouchard could not help but notice that many of the twins were dressed similarly.

Begun in the late 1970s, Bouchard's study continues to this day. Bouchard and his colleagues are eager to find out how alike in personality separated twins really are. From the data they collect, they hope to be able to shed some light on the age-old question: Is personality shaped mainly by heredity or by environment? One puckish sociologist answers that question with a simple "yes," to point up the fact that a strong ◊

case can be made either way.

The Minnesota project's results thus far suggest that heredity plays a much greater role in shaping personality, intelligence, mental health, and susceptibility to certain diseases than was previously suspected. Despite their markedly different upbringings, many of the identical twins studied display extraordinarily similar behavioral characteristics. Jack Yufe and Oskar Stöhr, for example, were separated shortly after their birth in Trinidad in 1933 and had met only once, briefly, in their early twenties. They were nearing fifty when Bouchard reunited them in 1979. Both came wearing moustaches, the same style of square wire-rim glasses, and blue, double-pocket shirts with epaulets.

Raised as a Catholic by his German mother and her family, Oskar had been a member of the Hitler Youth during the Nazi years. Jack was brought up by their Jewish father in Trinidad and later lived in Israel, where he worked on a kibbutz and served in the Israeli navy.

In spite of their dissimilar backgrounds, Jack and Oskar discovered that they shared a number of highly individual personality traits. For example, both said they enjoy sneezing loudly in elevators simply for the sake of watching people's reactions. Both read magazines from back to front, have formidable tempers, store rubber bands on their wrists, and habitually flush toilets before using them.

Striking behavioral similarities are seen in a number of other identical twins examined in the study. British twins Bridget Harrison and Dorothy Lowe are a case in point. Parted weeks after their birth in 1945, each came to the study wearing a watch and a bracelet on one wrist, two bracelets on the other wrist, and exactly seven rings on her fingers. Later, they learned that they both had owned cats named Tiger and that Dorothy had named her son Richard Andrew while Bridget had named hers Andrew Richard.

More intriguing, they had kept diaries for a single year when they were fifteen, then stopped. Each one bought the same brand, style, and color of diary. And although the entries themselves were not similar, both chose the same days to record their thoughts and the same days to omit them.

In Minneapolis, more than sixty sets of separated identical twins have supplied detailed medical and psychological information about themselves, responded to some 15,000 survey questions, and taken intelligence and perception tests. Many sets finished the tests at the same time, and they often made the same mistakes in filling in the answers. "The scores on so many of the tests are incredibly close," Bouchard commented, "closer than those of the same people taking the test twice."

The Minneapolis study also showed similarities in speech patterns, gestures, and movements among the twins, findings supported by European work. Related research in France, for example, has revealed that identical twins share comparable sleep patterns and experience corresponding cycles of dreaming. And a study at Rome's Gregor Mendel Institute suggests that twins can even develop similar ailments. As institute director Luigi Gedda put it, "If one has a cavity, the other has it in the same tooth, or soon will." □

THE CIRCUS OF THE BODY

Although every human body is unique, most fall within a norm—roughly standard in size, with the usual number of arms and legs, eyes and ears, fingers and toes. But dramatic anomalies are possible. People may be minuscule or gigantic, possessed of too many limbs or too few or none at all, or afflicted with deformities so grotesque that the sufferers hardly look human.

Among the "normal," tolerance for such radical extremes has never been strong. Fear and revulsion are among the more common emotions that anomalous humans evoke—along with a sort of morbid curiosity that has sometimes led to their unkind treatment as "freaks" to be displayed, gawked at, and scorned. The happy side of this cruel situation is that the anomalous have often made a virtue of their differences.

During the Middle Ages and the Renaissance, many human curiosities became prized possessions of aristocrats and royalty, serving as court jesters or valued retainers. Later, the more talented became self-employed impresarios.

The great age of oddities was the mid-nineteenth and early twentieth centuries. Circuses, sideshows, and exhibitions sprang up. Inhumane as they seemed, they did provide for the different ones a livelihood, a group of peers, a semblance of home. When sideshows began to lose their commercial luster in the 1930s, companionship and security became, ironically, both harder and easier to find.

The decline of the freak show was an indication that prejudice against the physically unusual was waning and that sympathy was on the rise. Some individuals who did not fit the norm found it easier to get commonplace jobs and live unremarked lives. But although society has become more tolerant of exterior peculiarities, few people would argue that it has become wholly accepting; and sympathy is very different from respect. The ordeal of being wildly different is not over.

England's Dashing Dwarf

Few men outside of fiction have lived a more celebrated and adventure-filled life than Jeffrey Hudson, a commoner-turned-courtier during the reign of England's King Charles I. By all accounts, Hudson owed his fame to his courage, to his good looks—and to his diminutive stature. When fully grown, he was about forty-five inches tall, the size of a typical five-year-old.

Hudson was born a butcher's son in 1619 in the town of Oakham. At age nine he was only eighteen inches tall, though perfectly formed. Early in life he joined the household of the wealthy and influential duke of Buckingham, who presented him to the king. The dwarf made his court debut by popping out of a cold baked pie.

Hudson's later career may well have grown outsize in legend, but the oft-told highlights are these: A favorite of Queen Henrietta Maria, the dwarf was once seeking a midwife for her majesty when he was seized by a Flemish pirate and held for ransom. His freedom bought, the dauntless courtier later took part in at least one foreign siege with the king's men.

In 1644, after Oliver Cromwell ousted the Stuart monarchy and installed his Protectorate in its stead, Hudson fled with the queen to France. There, five years later, he shot a man to death in a duel. Some say Hudson's antagonist had unwisely teased the dwarf about his well-known womanizing. After the duel Hudson ran away, was captured by Turkish pirates, and enslaved. Ransomed again, he was

back in England within ten years.

In 1679, some nineteen years after Charles II restored the monarchy, Hudson was briefly jailed on suspicion of joining an imaginary Catholic plot against the king. Once cleared and back in the Stuarts' good graces, the diminutive swashbuckler spent two years on Charles's payroll as a spy.

Hudson died in 1682, so famous that a suit of his clothes was sent to Oxford University's Ashmolean Museum—a reminder that minuscule size need be no barrier to mighty deeds. □

The Miniature Pimpernel

The Scarlet Pimpernel who saved aristocrats from death during the French Revolution was a fictional hero. But there was a real-life Pimpernel during the Reign of Terror, one who spirited life-and-death secret messages into and out of Paris for the aristocracy—feats he managed disguised as a baby.

His name was Richebourg, and he was a member of the household of the duchess of Orléans, whose son eventually became King Louis-Philippe. When the ominous tumbrels rolled through Paris on the way to the guillotine, Richebourg was a young man twenty or thirty years old and only twenty-three and one-half inches tall. The aristocrats desperately needed a foolproof courier to keep them in contact with allies outside Paris, and the dwarf bravely volunteered. A woman disguised as his nurse carried him past revolutionary strongholds wrapped in lace finery that concealed his vital dispatches. After making one of his forays, the cool secret agent was said to lean back in his baby clothes and nonchalantly light up a cigar.

Following the restoration of the monarchy, Richebourg was awarded 3,000 francs per year by the Orléans family in grateful recognition of his selfless derring-do. ☐

The Great Self-Promoter

In eighteenth-century Europe, a number of dwarfs thrived by ingratiating themselves with the nobility. So it was, for a time, with Josef Boruwlaski, who fell from grace with the aristocracy only to attain new heights as one of dwarfdom's most important self-promoters.

Boruwlaski was born in 1739 in Polish Galicia, where a local noblewoman, Countess Humiecka, took him into her household. The two toured Europe to some acclaim; in Vienna, Boruwlaski sat in the lap of Empress Maria Theresa, who gave him a ring from the hand of her daughter, Marie-Antoinette.

Witty and attractive, Boruwlaski called himself a count. All was well until at forty he fell in love with Isalina Barboutan, another Humiecka protégé. The jealous countess promptly threw him out. Boruwlaski appealed to Poland's King Stanislaus II for aid, but the stipend he got was not enough to support Isalina and the children that she soon bore.

It was then that the handsome dwarf hit on the idea of becoming his own exhibit, revisiting the noble homes he had stayed in with Countess Humiecka and sometimes staging piano concerts. He and his wife made their way through great houses in eastern and central Europe, although Boruwlaski sometimes complained of the ''torments'' that aristocratic ladies caused him with their caresses.

The couple's last stop was England, where King George III welcomed them warmly.

After the Boruwlaskis settled in Durham, England, money again became scarce. To make ends meet, the Polish impresario put himself on display at home. He died at ninety-eight and was buried in Durham Cathedral. ☐

America's Littlest Legend

The most famous dwarf in history was a cocky entertainer named Charles Sherwood Stratton, far better known as Tom Thumb. His renown was carefully nurtured by the showman Phineas T. Barnum. Together, the duo—one, five feet ten inches tall; the other, just under two feet when they met—were one of the greatest show-business teams the world has known.

At the peak of his glory, millions flocked to see tiny Tom perform, and European royalty begged to view him. Before his death at the age of forty-five, he had become an American legend.

Stratton was born on January 4, 1838, in Bridgeport, Connecticut, a carpenter's son. He weighed nine pounds two ounces at birth, and fifteen pounds two ounces five months later. But, although he continued to mature thereafter, Stratton's growth virtually ceased.

Barnum heard of the youngster in 1842 and hired him for a showing at his American Museum in New York City. The child and his mother arrived to discover that

Barnum was advertising "General Tom Thumb, a dwarf eleven years of age, just arrived from England." The name was an inspired choice: Tom Thumb was a popular nursery tale character, a tiny hero who performed miniature feats at King Arthur's court.

Barnum taught his own Tom Thumb to sing, dance, and impersonate such figures as Samson, Hercules, and Napoleon. Tom was, according to his mentor, "an apt pupil with a great deal of native talent and a keen sense of the ludicrous." With his own genius for advertising, Barnum crafted publicity for the new star, once plumping him down on a newspaper editor's dinner table to stroll along, shaking hands with guests. Every gimmick generated cascades of attention.

A national celebrity after only two years in the public eye, Tom was packed off for a three-year stint in Europe, where his fame hit new heights. He went three times to Buckingham Palace, where his outspoken wit captivated the young Queen Victoria. During Tom's first visit, the queen's poodle attacked him as he backed out of the room, and Tom fought off the animal with a cane. Such feats made him a favorite with the court, and Tom took on some of the trappings of a courtier himself. At Barnum's behest, the royal coachmaker built the diminutive celebrity a miniature carriage, only twenty inches high and eleven inches wide.

By 1857, Stratton was richer than Barnum, who had fallen on hard times. The loyal Tom went on a second European tour to save his partner from bankruptcy, and Barnum repaid him by introducing him to an engaging dwarf named Mercy Lavinia Warren Bump. When Tom's courtship of Lavinia proved successful, Barnum transformed the nuptials into one of the great media events of the era. Only 2,000 people were invited, and tickets were precious. Astors and Vanderbilts sent gifts, as did President and Mrs. Abraham Lincoln. The honeymooners stopped at the White House, where the lanky president declared that "God likes to do things in funny ways. Here you have the long and the short of it."

The wedding garnered so much publicity that Barnum cast about for a suitable follow-up. This came in the mid-1860s, when the showman announced that General Thumb and his wife had had a child. Then he hired a baby for Lavinia to hold during appearances. Later Barnum "revealed" that Tom and Lavinia's infant had died. Grief was lavish and widespread, and not until 1901 did Lavinia admit to the fraud.

In 1883, Tom Thumb died of a stroke. After his death it was learned that Stratton, who had earned millions on tour and in real estate, had squandered his wealth, leaving Lavinia with only $16,000 and some paltry property holdings. He was buried in Bridgeport, Connecticut, under a statue of himself that he had commissioned at the age of nineteen. It was life-size. □

Smallest of the Small

Lucia Zarate looked like a doll and weighed less than most house cats, traits that made her one of the most extraordinary dwarfs in history. Zarate, who was born in San Carlos, Mexico, in 1864, is said by experts to be the lightest person who ever lived. At maturity, the perfectly formed, doll-like woman was less than twenty inches tall and weighed about five pounds. Had she stood next to the adult Tom Thumb, she would have come to his elbow.

Lucia first appeared on exhibition in the United States at age twelve, and her pay reputedly rose to a heady twenty dollars per hour before her career ended in tragedy. Zarate was traveling through the Rockies during a snowstorm in 1890 when her train stalled. She died of exposure. □

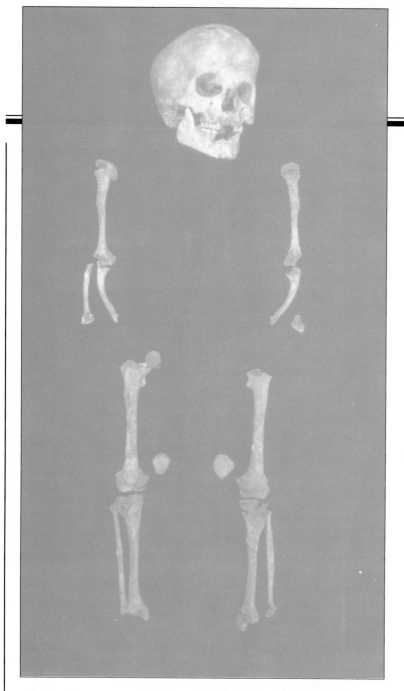

Bones of the Oldest Dwarf

The oldest evidence of human dwarfism is from an 11,150-year-old skeleton found in 1963 in southern Italy. Although remains from Paleolithic times reveal many misshapen individuals, nearly all of them were deformed by injury or by old age. The skeleton of the dwarf is the first example on record of a person who was obviously born deformed.

An examination of the bones shows that the individual was too weak to take part in the hunting and gathering that sustained his tribe. Nevertheless, he lived to the age of seventeen, indicating that he was cared for by his fellows. □

A Seal-Limbed Genius

Every now and then, extreme bodily peculiarities coexist with prodigious abilities. Such a case was Matthias Buchinger, an energetic, eerily talented, and, apparently, irresistibly attractive man who amazed Europe in the first half of the eighteenth century.

Buchinger was acclaimed as an artist, musician, magician, inventor, marksman, and athlete. He was one of the most renowned performers of any kind of his day. His attainments, unusual in themselves, were even more astonishing given that the self-exhibiting wonder stood only twenty-nine inches tall and had no recognizable arms or legs.

Buchinger, born in 1674 in Anspach, Germany, was phocomelic, a Greek term that means "seal-limbed." Out of his shoulders, where arms normally appear, protruded stubs ending in fingerless knobs. His legs were also stubs.

Fortunately, Matthias, youngest of nine children, had loving parents. He received a full education and was given the time to practice the skills that made his reputation later in life. Buchinger sought a living as a performer only when his father and mother died. His first recorded performance took place in Germany in 1709, when he was about thirty-five years old.

His skill with his flipperlike limbs was the product of tremendous patience and determination. Grasping implements with his two arm fins, he wrote and drew fluidly. His drawings, which he sold as part of his traveling show, were marvels, containing details that could not be seen with the naked eye. One self-portrait of Buchinger in a wig, when magnified, shows curls made up of the words of seven psalms and the Lord's Prayer.

A musician of professional caliber, he played the dulcimer, trumpet, bagpipe, guitar, oboe, flute, and drums. He gave musketry and fencing demonstrations and, as a magician, made objects appear and disappear. He was also an inventor, creating a mechanical device to do some of the disappearing tricks that other magicians effected with fancy fingerwork.

His disabilities notwithstanding, Buchinger was described by a contemporary as having been "very well shaped," with "a handsome face"—attributes not lost on several women. During his career he married four times and fathered at least eleven children.

Traveling back and forth across the Continent, Buchinger claimed at one time to have performed before "three Emperors, and most of the Kings and Princesses of Europe." Eventually he immigrated to England, where he found a special patron in Robert Harley, earl of Oxford, and drew the attention of King George I.

In his later years, Buchinger's fortunes faded, possibly from overexposure: Too many people had already seen his energetically marketed talents. He died in 1722. □

Three-Legged Inspiration

The eeriest extremes that the body can create often fall into the category of duplication—a designation covering deformities that would seem to belong only to science fiction. Extra limbs and even whole bodies can erupt out of the human form in bizarre arrangements, all resulting from prenatal defects. Those who suffer from duplication are often shunned as monsters. The most famous victim of the affliction, however, used his amazing condition to attain a happy and rewarding life.

Francesco "Frank" Lentini, born in 1889 in the town of Rosolini, Sicily, was normal above the waist. But below that point he had three legs, including one that extended from the base of his spine. Some early developmental mistake of nature was responsible for Lentini's condition. Either the extra leg was a malformation of a single fetus, or it was left over from a potential twin that failed to develop properly in the womb. In any case, the slightly shorter third leg was of no use for walking. It did, however, serve as a kind of portable stool.

According to promotional pamphlets churned out during Lentini's later years, his father exiled him from home. The boy grew up withdrawn and depressed, the story goes, but his life changed when he visited a home for the handicapped. There he saw children who were blind, deaf, and even more deformed than himself, and the incident gave him a new appreciation for his life and a desire to improve it.

Lentini immigrated to America in 1898 at the age of nine and joined the circus. He appeared with Buffalo Bill's Wild West Show, Ringling Brothers, and Barnum & Bailey, among others, billed as the Three-Legged Wonder.

A handsome and active man, Lentini married and fathered four normal children. He retired to Florida, where he died in 1966. □

Germany's Armless Wonder

People born with vestigial arms were common attractions in nineteenth-century sideshows, but there was nothing common about Carl Hermann Unthan. By general agreement, no curiosity of the day was more gifted—or more determined—than Unthan, a Prussian schoolmaster's son who confounded audiences as a virtuoso violinist who played using only his feet.

Unthan was born in 1848 with stumps instead of arms, each protrusion bearing a tiny finger. His father forbade any coddling of the boy, and young Carl responded with an iron-willed persistence to overcome his plight. At two, he began feeding himself, using his toes to hold a spoon. Around the age of ten, he taught himself to play the violin by strapping the instrument onto a stool, then bowing with one foot and "fingering" the strings with the other. Some ten years later, he gave a debut concert performance that launched a half-century vaudeville career.

On stage, he was a marvel of resourcefulness and aplomb. When he broke a violin string during a performance on his maiden tour, he replaced it with his toes before the mesmerized audience. (He was later said to break a string on purpose as part of his act so he could repeat the impressive maneuver.) An accomplished marksman, he could shoot the spots out of a playing card with a rifle that he operated with his foot. During his long career, he performed throughout Europe and in the United States, Cuba, Mexico, and several South American countries.

With typical Prussian precision, Unthan called his 1925 autobiography *The Pediscript,* distinguishing it from a manuscript. Before he died in 1928, German officialdom recognized the Armless Wonder, as he was called, with a circus benefit in his honor. □

An Odd Pair

Charles Tripp and Eli Bowen, who toured with P. T. Barnum, had opposite afflictions, but both used their bodily peculiarities in successful circus careers. Tripp, a Canadian, was an armless cabinet-maker, while Bowen, from Ohio, was a legless acrobat.

Born in 1855, Tripp learned to feed, dress, and even shave himself with his powerful toes. He became a skilled woodworker before signing with Barnum in 1872. Known as the handyman of the circus, he toured for more than fifty years, demonstrating his skills at carpentry, penmanship, portrait painting, and ornamental paper cutting.

Bowen's feet grew directly out of his hip joints. He became a tumbler, swinging acrobatically from a pole. Married, with four children, he later retired to California. □

The Happy Half Boy

Sideshow attractions often displayed themselves on a platform built of exaggeration and ballyhoo—so much so that it was often hard to separate the genuine achievements of these touring oddities from the fantastic claims that they handed out in brochures to gullible audiences.

Certainly, the aroma of show-business hype clung heavily to the publications that promoted Johnny Eck, described as "The Only Living Half Boy: Nature's Greatest Mistake." But there was no mistake about either the startling nature of Eck's deformities or the unique combination of optimism, courage, and humor that made him a favorite of American audiences in the Roaring Twenties.

Whether he was the only living

example of his type is at least open to question, but the description of Eck as a half boy was sober truth. He was normal from the waist up—and nonexistent below the rib cage. Vital organs below that point were simply compressed upward into his trunk. When he began touring at fourteen, he stood only eighteen inches tall and weighed fifty-seven pounds. Yet he conquered his disabilities to become a fine swimmer, diver, acrobat, and juggler who performed a high-wire act. He was also a magician and a comic. In one of his most spectacular achievements, Eck performed as a dancer—on his multitalented hands.

Eck's personality was as striking as his physique and his talents. Despite his great handicap, he struck those who knew him as extraordinarily charming and cheerful. He had a lifelong fondness for children—and they for him—perhaps because they seemed better equipped than adults to accept him without regard for his deformities. "It is strange that children feel such innocent love for me," he once said.

Eck's real name was John Eckhardt, and he was born in Baltimore in 1911, along with an identical twin who was perfectly formed. According to an autobiographical brochure, he learned to walk at the age of one by pulling himself along with his hands. He attended grade school and high school with his brother, and by his own account—which was probably embroidered—Eck attended a Baltimore business college, graduating with "the highest honors ever given to anyone."

Young Johnny began performing at age fourteen and was soon touring the United States and Canada with his twin. (In the off-season, Eck claimed—although some people are skeptical—that he composed for and conducted a twelve-piece orchestra back in Baltimore.) Maestro or not, the Half Boy was noted for his sartorial elegance. He acquired a special taste for luxurious, custom-made silk shirts.

The half man eventually came to the attention of Hollywood. He landed a part in a Tarzan movie, although his performance as a giant bird ended up on the cutting-room floor. Nevertheless, he went on to gain some fame for his appearance in the controversial circus movie *Freaks*, produced by Metro-Goldwyn-Mayer.

Eventually, Eck's exhibition career went into decline, and he and his brother became part owners of a traveling penny arcade. Today, the brothers, now retired, live together in the Baltimore house where they were born. □

The Exploited Apewoman

Publicity pamphlets of the last century described Julia Pastrana as being of "semihuman" stock. One physician declared her a hybrid of a human being and an ape, while another doctor proclaimed her to be a "distinct species."

The overblown claims badly distorted the truth about a Mexican woman whose strange nature needed no further distortion. Billed as "the ugliest woman who ever lived," Pastrana was only four and a half feet tall, and her slight stature was easily the least noticeable anomaly of her appearance. She had a dark, curly beard that hung under her chin. Thick hair covered much of the rest of her face and her body. Heavy browridges thrust above her flat nose, while her jaws pushed forward like a gorilla's, further exaggerating lips that ◊

were already twice normal size.

Modern medicine would come to define the source of Pastrana's hairiness as a hormone imbalance, possibly an overproduction of male sex hormones by her ovaries. Her apelike face may have had the same cause, while her outsize lips were backed by unusually thick gums, possibly the result of vitamin C deficiency.

Pastrana first appeared onstage in America in 1854, and she toured the United States and Europe for about five years. Then she married her manager, who evidently feared that his star might leave him. In 1860, she died in Moscow of complications following the birth of her only child. Her hus-band—apparently no wellspring of sensitivity—embalmed her body *(right)* and that of her short-lived son, and displayed them in a glass case before anyone willing to pay to see them. The ghoulish husband died in 1884, but the grotesque exhibition passed from hand to hand across Europe for decades. The corpses returned to America in 1972 on a final tour before recrossing the Atlantic and vanishing in Norway. □

The New York Public Library

The Gentle Giant

Robert Wadlow was determined not to be set apart as a freak. His dignified insistence softened the fact that his goal was plainly impossible: Before he died at age twenty-two, Wadlow stood a little over eight feet eleven inches tall and weighed 491 pounds. Medical science considers him the tallest man who ever lived.

Wadlow weighed just eight pounds six ounces when he was born in Alton, Illinois, in 1918. But by the time he was five, he stood five feet four inches tall, and at eight was six feet two and one-half inches—taller and stronger than his father. Doctors diagnosed his giantism as acromegaly, a condition caused by overactive pituitary glands. The hormonal imbalance was never checked because researchers apparently felt that a brain operation to correct the problem might kill him.

Wadlow's caring parents encouraged their son to lead as normal a life as ◊

Shown actual size is one of the custom-made shoes worn by the young giant Robert Wadlow, who towers over his family in the picture above.

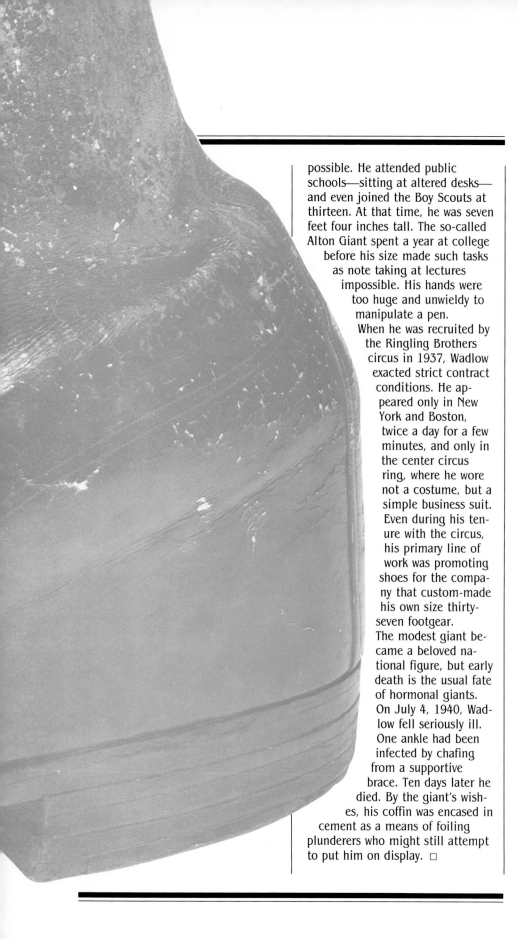

possible. He attended public schools—sitting at altered desks—and even joined the Boy Scouts at thirteen. At that time, he was seven feet four inches tall. The so-called Alton Giant spent a year at college before his size made such tasks as note taking at lectures impossible. His hands were too huge and unwieldy to manipulate a pen.

When he was recruited by the Ringling Brothers circus in 1937, Wadlow exacted strict contract conditions. He appeared only in New York and Boston, twice a day for a few minutes, and only in the center circus ring, where he wore not a costume, but a simple business suit. Even during his tenure with the circus, his primary line of work was promoting shoes for the company that custom-made his own size thirty-seven footgear.

The modest giant became a beloved national figure, but early death is the usual fate of hormonal giants. On July 4, 1940, Wadlow fell seriously ill. One ankle had been infected by chafing from a supportive brace. Ten days later he died. By the giant's wishes, his coffin was encased in cement as a means of foiling plunderers who might still attempt to put him on display. □

SENSES

Every living creature interacts constantly with the world around it, foraging and mating, fleeing and fighting, navigating an environment fraught with obstacles. The chief instruments of such interaction are the senses, which gather data from outside the body and present it to the brain. Like most animals, human beings possess five senses—those of hearing, vision, taste, smell, and touch—each one with its own organ of perception. The eyes see, the ears hear, the tongue tastes, the nose smells, and the skin feels; but, in reality, very little happens until their signals are collected and interpreted inside the brain.

Although finely tuned to certain ranges of sensitivity, human senses can be both better and worse than the supposed norm. Some people can hear a broad range of sounds with remarkable sharpness, for example, while others cannot hear at all. A few have eyes as keen as hawks', but many have lost, or never had, their sight.

When one sense fails, the others can be trained to compensate. The hearing of the blind sharpens; the deaf develop an eye for nuance and detail. Such mutual reinforcement demonstrates that, though separate in some respects, the five senses are in fact a composite of capabilities.

Window on the World

The eye is the most sensitive of all the human sensory organs. Approximately 70 percent of all the body's sense receptors are found in the retina, a screenlike surface on which incoming light is concentrated. In the retina, this complex light-sensing apparatus is packed into an area only about a tenth of an inch in diameter.

A large bundle of nerves carries the signals from each retina to the primary visual cortex, the part of the brain that tells us what we see. Each one of these optic pathways contains hundreds of thousands of nerves.

The retina contains two types of light-sensing cells. Seven million or so cone cells detect color, giving the human eye its remarkable ability to distinguish colors—some seven million shades. More than 100 million rod cells are used mainly to see in low light. Because these sense only black and white, the world we see loses its color as it loses its light. Rod cells can sense approximately ten million shades of gray and can detect as little as 100 trillionths of a watt, equivalent to the flame of a cigarette lighter seen from fifty miles away. On the retina, such a point

of light would produce an image less than one nine-thousandth of an inch in diameter, about the size of a red blood cell.

Vision gets its sharpness partly from the rod cells and partly from the muscles that move the eyes and make tiny focusing adjustments in the optical shape of the organ. These muscles move about 200,000 times a day. The sharpest eyesight known belongs to Veronica Seider, a dentist from Stuttgart, West Germany. With vision some twenty times better than the average, she can identify people at a distance of more than a mile. □

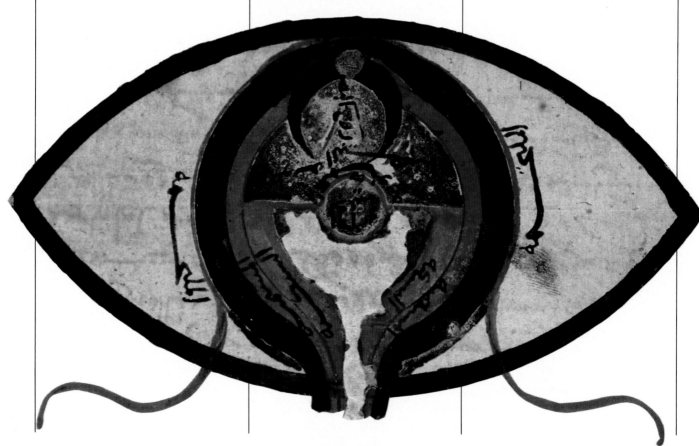

Recalled to Light

In fiction, the drama of surgically restored sight is a familiar one: The bandages come off and the patient cries, "I can see!" Certainly there is great joy in the recovery of vision, but the experience can also be terribly disorienting to those who have never seen before.

Studies of the newly sighted suggest that they find themselves in a world where nothing is quite right. Is a warm shade of color a red or a pink? Is a portrait the same as a real person? Is a distant mountain the same size as a nearby hill? Trained to the language of sounds, odors, tastes, and feels, their vocabularies fail them in the glare of new vision, and even the most familiar things become temporarily strange.

In one classic example, a teenager, blind so long that he had no memory of seeing, took months to understand his transformed world after his vision had been restored. The things he knew by shape, texture, weight, and smell had taken remarkable forms that were difficult for him to remember. His pet dog and cat, for example, were hard to identify correctly unless he could touch them.

The human eye sees colors with particular sharpness. But for a person blind since birth or infancy, distinguishing between shades and intensities of color, or reconciling the fact that two objects are identical except for their color, can be quite difficult.

The newly sighted must also learn how to deal with the concepts of size and perspective. When they first begin to see, they report that objects seem to be touching their eyes and that they have to force what they see back into perspective. As a result, everything seems incredibly large. Although this distortion soon goes away, other problems linger. For example, the teenager who had trouble telling his cat and dog apart without touching them also had problems accepting the substitution of images for the real thing. Shown a locket-size portrait of his father, he recognized the likeness but was not immediately convinced that something the size of a human face could be rendered as small as a half dollar.

For some, renewed sight is a burden rather than a gift. Formerly blind people may be disturbed when they realize that others can observe them without having to touch them—that the sighted world is full of people watching others without their knowing it. One man, cheerful and productive while blind, became depressed soon after gaining sight. Such blemishes as flaking paint disturbed him greatly, and he soon found the visual world drab. Evenings were particularly dispiriting, for the bright colors he enjoyed faded with the day. In the end, the operation that was intended to enrich his life brought only depression. □

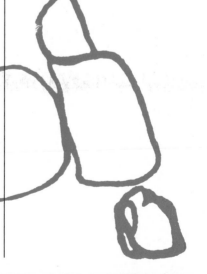

Four months after a sight-restoring operation, a newly sighted patient's drawings of a person (upper left), a house (center), and an automobile (bottom) show little evidence of detail or perspective.

Facial Vision

In 1749, the French philosopher Denis Diderot noted "the amazing ability of a blind acquaintance not only to perceive the presence of objects, but to judge accurately their distance from him." Diderot concluded that this extraordinary spatial sense was caused by an increased sensitivity of the nerves in the man's face.

Over the next century, many claims were made for the existence of something called facial vision. One scientist even theorized that blind people used their cheeks as feelers. In 1905, another Frenchman, Émile Javal, coined the term *sixth sense* and claimed that it was aroused as the face was struck by waves of ether, an invisible, universal medium once wrongly believed to fill space.

Not until 1944 was the supposed existence of the sixth sense systematically explored. Then, in a series of experiments conducted at Cornell University in Ithaca, New York, psychologists Michael Supa, Milton Cotzin, and Karl Dallenbach tested the ability of blind students and blindfolded sighted students to detect obstacles placed in their paths. With a little practice, the blindfolded subjects became as good as their blind colleagues at perceiving and avoiding obstacles.

To test the notion of facial vision, the researchers covered the subjects' faces with heavy felt blanket material and put the students through the obstacle test again. Despite being deprived of a sense of feeling, both sighted and blind subjects were able to perceive obstacles and avoid them.

Both groups did poorly, however, when they walked in their stocking feet over thick carpeting, wore earplugs, or had their hearing "jammed" by loud background noises. Even though their faces were open to stimulation, they failed in every one of hundreds of tests in which their hearing was somehow impeded.

Besides debunking the myth of facial vision, the Cornell scientists had made an important discovery about how the blind navigate. By paying particular attention to sounds and how they subtly change in different surroundings, the visually impaired develop a human counterpart to the echo-location techniques of bats, porpoises, and whales. Some blind people even use muted clicking noises made with the tongue and listen to the extremely faint echoes returned from the objects around them. One man who uses a cane to help him navigate says that he can tell when he is approaching an obstacle by changes he hears in the tapping noises from his stick.

Although it seems that people who have lost the use of their eye-sight develop a keener sense of hearing, the fact is that they learn how better to discriminate between various sounds. Thus it is listening, not hearing, that sharpens with blindness. □

Foolish Eyes

Sharp as it is, human vision is easily tricked. Because visual information is interpreted by the brain, we see what we think we see. The perceived image is constructed from incoming light signals and past experience, which adds another layer of information to make a composite of imagination and reality. It is often wrong.

In the so-called Ponzo illusion, shown at right, the two horizontal lines appear to be of different sizes. To the brain, converging lines mean distance, whether on a page or in the form of iron rails stretching to the horizon. As far as the brain is concerned, the top horizontal line must be farther away and physically longer than the bottom one. But the Ponzo diagram is just a trick. The horizontal lines are the same size.

One of the great physical illusions for the eye was created in the 1950s by psychologist Adelbert Ames, Jr. In the disorienting Ames Room *(above)*, two people of about the same size seem to be of vastly different sizes. Again, the brain colors what it sees with what it knows. Based on experience, it assumes rooms with vertical walls and corners are rectangular, therefore the girl and boy shown here should be about the same distance from the eye. But the room is actually a trapezoid, in which the rear right corner is closer to the viewer than the rear left corner, the left-hand window is larger than the right-hand one, and the seemingly flat ceiling slopes down from left to right. The eye may see the trick, but the brain, blind to the illusion, perceives a giant boy. □

Dreams of the Blind

Most people are bombarded with visual images throughout their waking lives. They know the world as an endless tapestry of forms and faces. During sleep, this visual barrage translates itself into the kaleidoscopic pictures that compose dreams. Psychologists have generally regarded dreaming as an almost exclusively visual experience. But this suggests that only those who see, or have seen, can dream. In fact, the blind have dreams that are in their own way as vivid as anything that is experienced by the sighted.

Blind people interpret their daytime world through hearing, touch, taste, and smell, and through the senses of balance and motion. Those who were born blind or lost their sight in early childhood have no concept of seeing with the eyes. Accordingly, their dreams are rich in sound, movement, and tactual images, and visual pictures are altogether absent.

In a study of sixty-seven blind Illinois schoolchildren conducted in the 1920s, researcher Elinor Deutsch found that hearing was the most important mode of perception, followed by touch. One girl dreamed that she was in the middle of a herd of buffalo, which she recognized by their stamping and bellowing. Another dreamed that her mother suddenly fell to pieces, her bones rattling across the kitchen floor.

Elinor Deutsch, herself blind from birth, reported one of her own dreams, in which she was having dinner with her sister, brother-in-law, and their two children. At the table an argument flared between her sister and brother-in-law over the purchase of some art supplies for one of the children. After an exchange of angry words, her brother-in-law uttered a fierce, guttural cry and sprang up, brushing Deutsch with his outstretched arm as he lunged at her sister. Before actually striking her, he hesitated, then left the room. Deutsch remained seated at the table, listening to the agitated breathing of her sister and the whimpers of the children.

This dream is recounted almost entirely in terms of auditory and tactual perceptions. The dreamer hears the argument, the snarl of rage, the sounds of breathing and frightened youngsters. Feeling her brother-in-law's extended arm, Deutsch knows he is about to strike her sister. Although none of this episode of family violence was seen, it emerges fully rendered in Deutsch's dream.

Collaborating on a series of studies in the 1970s, Donald Kirtley of California's Fresno State University found that blind people, like people who can see, dream about objects and activities that are prominent in their daytime environment. Blind people who travel on the job dream about airports and taxis, while those whose lives are sedentary dream of familiar rooms at home. And there are the usual human anxieties and hopes—lost keys, exams not studied for, winning the lottery.

Interest in the dream imagery of the blind dates back to the nineteenth century. In 1838, a German investigator named H. Heermann pinpointed the time between ages five and seven as a critical period for the retention of visual dreams. People in his study who had lost their sight before the age of five never dreamed in visual images. Those who had vision until at least the age of seven continued to see in their dreams, in some cases for as long as fifty years. But most of those who lost their sight between ages five and seven gradually lost the visual component of their dreams, which became increasingly the hearing and feeling dreams of the always blind. Subsequent research has tended to confirm these early findings.

In 1955, another German researcher, H. von Schumann, applied Heermann's work to a study of dreams described in the *Iliad* and the *Odyssey* of Homer, who, according to legend, was blind. Because the Greek poet told of dreams consisting almost entirely of auditory and tactual cues, and little or no visual imagery, von Schumann concluded that Homer was blind and that he must have lost his sight very early in life. □

The Secret Audience

Physicians once assumed that patients under anesthesia were totally unconscious and unaware of what was discussed in the operating room. There is growing evidence, however, that the sleeping subjects can hear rather well—and that they often form memories of such discussions.

In a series of experiments conducted in the 1980s, University of California psychologist Henry Bennett told patients under anesthesia that they should tug their ears when interviewed after surgery. In later interviews, the patients tugged their ears an average of six times, even though they reported no conscious memory of being told to do so.

Some doctors believe they can use this subconscious awareness to help patients through difficult surgery and to faster recovery. Carlton Evans, a medical researcher at London's St. Thomas Hospital, told some anesthetized patients things such as "you'll want to get up and get out of bed to help your body recover earlier." The patients given such positive messages did indeed heal more quickly and with fewer complications than those who received no instructions.

The power of such suggestion has a downside, however. Glum prognoses or disparaging remarks about the patient uttered in the operating theater may translate into subsequent depression and difficult convalescence. Totally innocuous statements during surgery can also cause trouble. A woman being operated on for cancer heard her surgeon say, "I can't get it all out." When she came out of anesthesia, she could not shake a feeling of depression. Finally, eighteen months later, she confronted her physician with what she had heard. The shaken surgeon explained that he had been talking about algae in his pool. □

So Sensitive

Sound is what the ear and brain make of vibrations that ripple out from a source, transmitted from molecule to molecule in any medium but a vacuum—in which sound cannot exist. The ear senses these vibrations and passes them to the brain for interpretation. Although human hearing is less acute than that of some animals, it is in fact remarkably sensitive to sound waves over a considerable range.

We can detect sound vibrations ranging in frequency from about 20 to 20,000 vibrations, or cycles, per second. These include sounds lower than a piano's lowest note and higher than a piccolo's highest note. Noise lower than 20 vibrations per second is infrasound; higher than 20,000 cycles lies ultrasound. Humans hear neither, but many animals do. A German shepherd can hear sounds as low as 15 cycles per second, the infrasonic levels at which, according to some naturalists, elephants communicate. Fruit bats and vampire bats can hear pitches of 120,000 cycles per second, while dolphins detect sounds as high as 150,000.

Though not the best, the human ear offers more hearing ability than most people are able to use. Remarkably versatile, our listening machinery can hear meadow grass blowing in a gentle breeze but still tolerate the sonic blasts of a jackhammer, some ten trillion times louder. And the ear is selectively tuned to be more sensitive to higher pitches than lower ones, which is why a woman's voice can be heard farther away than a man's. Among other things, this keeps the ear from being bombarded by unwanted bodily noises, most of which are at very low frequencies. One potentially devastating problem is solved by an absence of blood vessels where the ear converts vibrations to nerve signals. If blood circulated there, all that we could hear would be a deafening roar of pulse. □

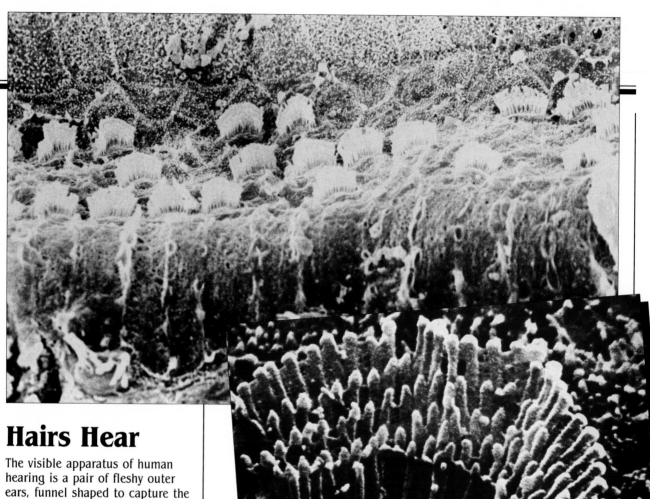

Hairs Hear

The visible apparatus of human hearing is a pair of fleshy outer ears, funnel shaped to capture the vibrations that can be heard as sounds. Brought in through a narrow, circuitous canal, these vibrations are picked up and imitated by a taut tissue, the eardrum. It passes the signals to the three tiniest bones in the body, the hammer, anvil, and stirrup, which relay the sound through the thin membrane of a passage called the oval window into the fluid-filled spiral of the inner ear, or cochlea. Transformed into gentle waves, the sound vibrations finally reach a thin tube, the organ of Corti, suspended in the middle of the cochlea. Here, sound becomes nerve signals that travel to the brain.

Inside the organ of Corti are some 25,000 tiny hairlike cells, known as cilia, that wave back and forth like patches of sea grass stirred by a gentle ocean current *(above)*. The listening hairs are grouped in small semicircles connected to yet another membrane that generates a nerve signal whenever a cilium moves.

Since high-frequency vibrations are closely spaced, they need to be detected near the entrance to the organ of Corti. Accordingly, the cilia closest to the oval window are tuned to sense the highest audible frequencies, around 20,000 vibrations, or cycles, per second. Those at the far end of the organ of Corti are sensitive to low-frequency vibrations, corresponding to tones around 20 cycles per second. Within the range of human hearing, cilia are sensitive to even the slightest vibration, which enables ears to hear a pin drop. But their sensitivity also means that they can be damaged when the vibrations are too strong. Once hurt by exposure to a loud noise, cilia do not heal; hearing lost this way is lost forever. □

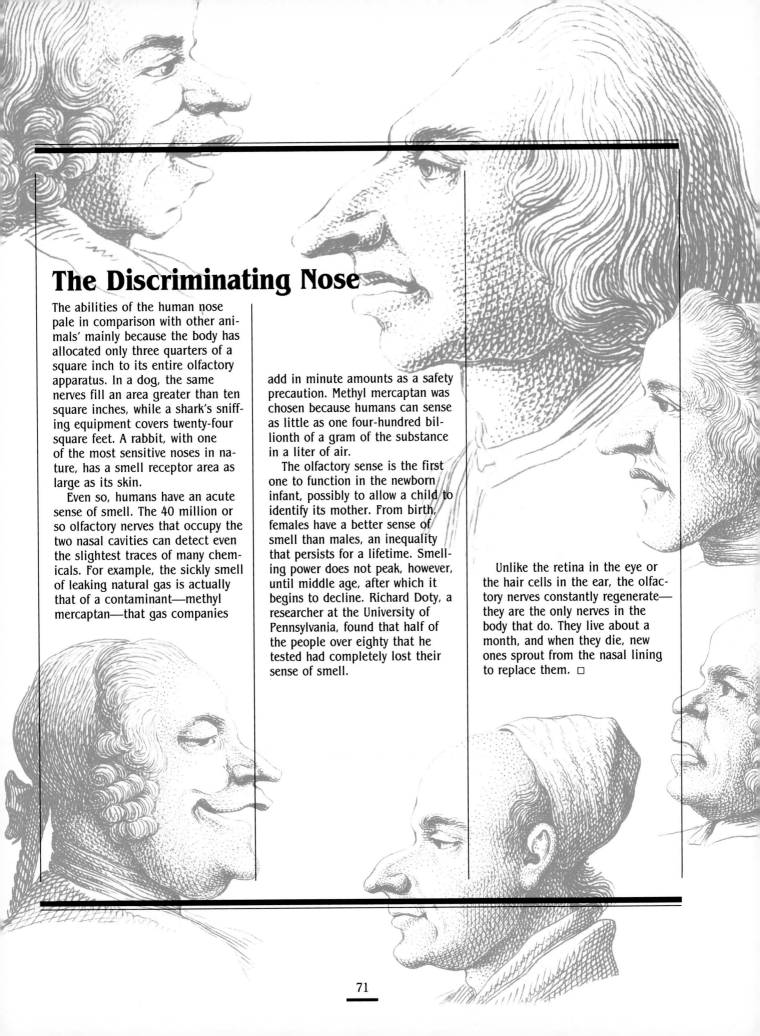

The Discriminating Nose

The abilities of the human nose pale in comparison with other animals' mainly because the body has allocated only three quarters of a square inch to its entire olfactory apparatus. In a dog, the same nerves fill an area greater than ten square inches, while a shark's sniffing equipment covers twenty-four square feet. A rabbit, with one of the most sensitive noses in nature, has a smell receptor area as large as its skin.

Even so, humans have an acute sense of smell. The 40 million or so olfactory nerves that occupy the two nasal cavities can detect even the slightest traces of many chemicals. For example, the sickly smell of leaking natural gas is actually that of a contaminant—methyl mercaptan—that gas companies add in minute amounts as a safety precaution. Methyl mercaptan was chosen because humans can sense as little as one four-hundred billionth of a gram of the substance in a liter of air.

The olfactory sense is the first one to function in the newborn infant, possibly to allow a child to identify its mother. From birth, females have a better sense of smell than males, an inequality that persists for a lifetime. Smelling power does not peak, however, until middle age, after which it begins to decline. Richard Doty, a researcher at the University of Pennsylvania, found that half of the people over eighty that he tested had completely lost their sense of smell.

Unlike the retina in the eye or the hair cells in the ear, the olfactory nerves constantly regenerate—they are the only nerves in the body that do. They live about a month, and when they die, new ones sprout from the nasal lining to replace them. □

Remembrance of Smells Past

Of all the senses, smell may be the most powerfully evocative. A whiff of apple pie or burning leaves, for example, can trigger a stream of long-forgotten memories from early childhood. One man, deprived of his sense of smell when injured in a car accident, said he mainly missed the memories set off by a familiar odor. In a way, his inability to smell cut him off from part of his past.

But smell is also a kind of primal sense, evoking not the recent but the very ancient human past. The most primitive of the senses, smell is the ability possessed by almost every organism to detect the chemistry around it. Indeed, the sense of smell seems to have developed very early in human evolution—some say the brain itself evolved from an olfactory organ.

The heart of the olfactory system—two small bulbs—lies deep within the brain. The bulbs receive information from 40 million olfactory nerves that hang from the roofs of two nasal cavities. Although these nerves connect directly to the olfactory bulbs, they are shielded from the outside world only by a thin coating of mucus. They are the brain's most direct link to physical environment, perhaps a remnant of organs that first developed some 500 million years ago.

Surrounding the olfactory bulbs is the limbic region of the brain, which controls emotion and plays a pivotal role in forming and recalling memories. This intimate association explains why odors and emotions are linked. Indeed, Yale University researchers have found that specific odors can elicit particular emotions in test subjects: An orangelike scent, for example, made people feel relaxed, while lavender was stimulating; a whiff of spiced apple reduced anxiety and even lowered blood pressure.

The sense of smell may also play a role in how humans interact with one another. The limbic system is connected to the hypothalamus, the master control center for many of the body's functions *(page 13)*. Pheromones, chemicals the body emits in amounts too small to be smelled but still detectable by the olfactory system, appear to trigger interactions between these glands, with sometimes surprising effects.

For example, it is a well-established fact that women who live and work in close proximity, as in dormitories, soon acquire synchronized menstrual cycles. Researchers at the Monell Chemical Senses Institute have produced the same effect merely by placing small amounts of underarm sweat from one group of women under the noses of a second group of women. Presumably, the olfactory-limbic-hypothalamus combination did the rest.

Scientists have found that almost all animals emit pheromones to attract mates, identify family members, and warn off enemies. Thus, when two people are said to share a special chemistry, perhaps they really do. □

Sippers

Nine stubby wine glasses are ranked on a plain wooden table; behind them stand nine bottles of red wine, their labels covered with foil. The winetaster uncorks the bottles, rinses each glass with wine from the bottle behind it, then pours an inch or so of the rich, red liquid from the first bottle into the first glass. He holds the glass up to the light in order to check the wine's color, swirls the wine, and inhales its fumes. After remarking on the wine's nose, or bouquet—it may be woody, fruity, jammy, mushroomy, dumb—the taster finally sips the wine, sloshing it around in his mouth for a moment before spitting it expertly into a nearby sink.

It all looks like work for extraordinarily sensitive noses and tongues, and winetasters enjoy a reputation for keen senses. What they actually possess, however, is a remarkable knowledge of the various flavors that wines can have and a vocabulary with which to express them. For example, "foxy" refers to a distinctive earthy tang and flavor of wine made from native American grapes. On the other hand, "goaty" wine has a rich, ripe animal-like flavor.

But tales of wine gurus who can tell everything about a wine with but a whiff and a gargle are highly exaggerated. A recent proof of this came from a 1985 "sipstakes" in California, where only 1 of 250 tasters in this annual event could correctly name the year, county, and producer of a wine. □

Like a garden of pink desert plants, large taste buds called vallate papillae *(center)* grow against lymphatic tissue *(left)* at the base of the tongue, flanked by smaller, cactus-shaped filiform and fungiform papillae *(right).*

A Matter of Taste

The so-called taste of food actually refers to its flavor, which is a combination of three qualities—aroma, temperature, and texture—plus a fourth provided by the sense of taste itself. Although this sense has its detectors, or buds, in the mouth, flavor is some 80 percent aroma, detected by the nose. To prove the relationship, one need only squeeze the nostrils together before biting into two different-tasting foods. An apple and a raw potato, for example, will both taste slightly sweet when there is no aroma to guide the brain, as will an orange and a grapefruit.

The actual sense of taste detects sweetness, saltiness, sourness, and bitterness, the four components of taste. These qualities are identified by tiny taste buds, located mainly on the tongue, but also found throughout the mouth and even in the throat. The buds for detecting sour and bitter substances are about 10,000 times more sensitive than the sweet buds, which are the least discriminating. This sharpened sensitivity to bitterness may have helped early humans survive—many poisonous plants have a repellently bitter taste and would have been avoided as food.

Like the other senses, taste often grows dull with age. Indeed, many elderly people have lost so many taste buds that they lose interest in food, thus depriving themselves of needed nutrients. □

Unwell Smell

Many diseases are known to produce characteristic odors on the bodies of the afflicted: Typhoid smells like baking bread, German measles like plucked feathers, yellow fever like a butcher shop, and gangrene like a rotten apple. Once, sensory analyst David Kendall was asked by doctors at Boston's Massachusetts General Hospital to help with a perplexing case involving a comatose child who had a strange odor. After sniffing the patient and conferring with physicians, Kendall was able to help pinpoint the child's illness, which turned out to be a rare metabolic disease linked to certain foods. With an appropriate change of diet, the child quickly improved. □

Million Dollar Noses

Every human endeavor has its specialists, and supplying smells and tastes for profit is no exception. Someone must determine how good dog food tastes or whether an underarm deodorant does its job. In the trade, these specialists are called sensory analysts, people who can train their senses to detect odors and flavors of all types.

Perhaps the most famous in the field was the late Ernest Charlton Crocker, who according to legend could discriminate between more than 9,000 different odors. In 1951, he was dubbed the "man with the million dollar nose" when, by sniffing various concoctions, he was able to come up with a new formula for Lux detergent that smelled like the original but was cheaper to produce.

The remarkable sensitivity of sensory analysts is not inherited, but acquired through training that turns a normally functioning person into a professional sniffer and taster. Most companies get their sensory analysts from their pool of employees. One firm, for example, regularly tests all of its new employees for their smell and taste abilities. At the Arthur D. Little

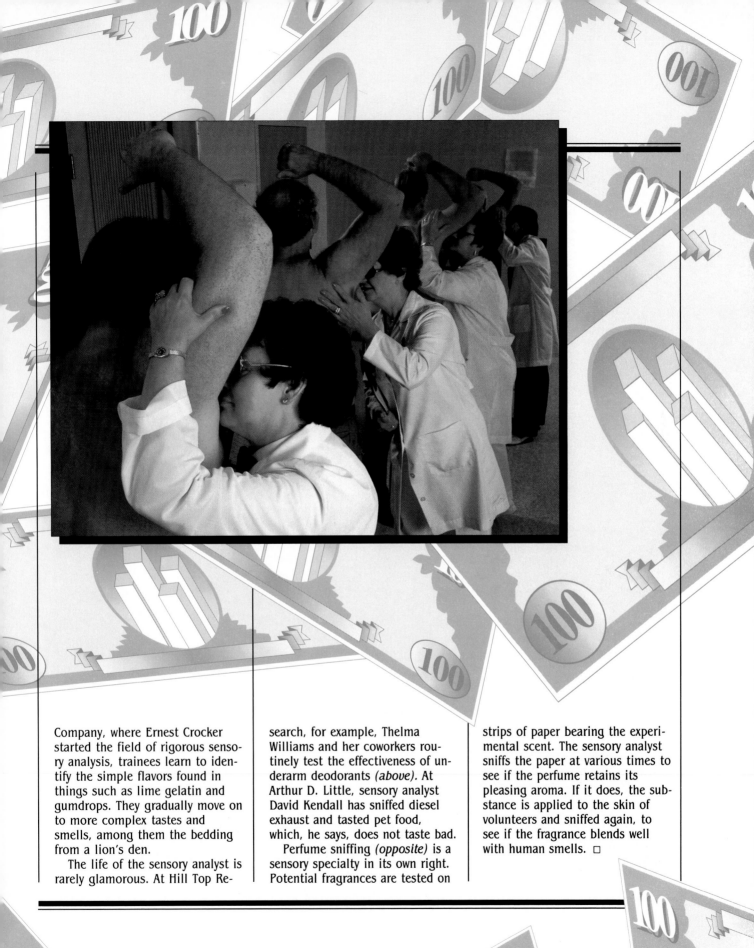

Company, where Ernest Crocker started the field of rigorous sensory analysis, trainees learn to identify the simple flavors found in things such as lime gelatin and gumdrops. They gradually move on to more complex tastes and smells, among them the bedding from a lion's den.

The life of the sensory analyst is rarely glamorous. At Hill Top Re-search, for example, Thelma Williams and her coworkers routinely test the effectiveness of underarm deodorants *(above)*. At Arthur D. Little, sensory analyst David Kendall has sniffed diesel exhaust and tasted pet food, which, he says, does not taste bad.

Perfume sniffing *(opposite)* is a sensory specialty in its own right. Potential fragrances are tested on strips of paper bearing the experimental scent. The sensory analyst sniffs the paper at various times to see if the perfume retains its pleasing aroma. If it does, the substance is applied to the skin of volunteers and sniffed again, to see if the fragrance blends well with human smells. □

Touching

Like taste, the sense of touch is a composite one, comprising at least eight different stimulations: heat, cold, movement, delicate movement, vibration, steady contact, stroking movement, and pain. Researchers believe that there may be many more that have not yet been detected.

Receptors *(right)* tailored to such stimuli are found over the entire body, but not in a uniform distribution. For example, the receptors for delicate movement are found only in the base of hair follicles. The fingertips, lips, and tongue contain the greatest density of touch sensors, while the skin of the back has relatively few.

The sense of touch has many special emotional qualities. In many cultures, hugging, kissing, a pat on the back, and a warm handshake all produce a positive feeling. Researchers have even found that touching is essential for life: Newborn babies who are not touched do not thrive and sometimes even die from a condition called marasmus. At New York's Bellevue Hospital, infant mortality sharply declined after nurses began a daily regimen of holding and stroking new babies. □

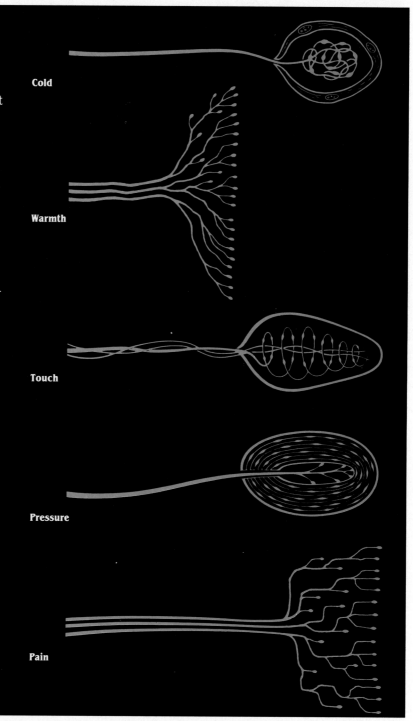

Cold

Warmth

Touch

Pressure

Pain

A Sensory Enigma

Sarah, as the seven-year-old girl was called, had never felt pain. Although this might have seemed a blessing, it was actually quite dangerous, for pain is the way the body identifies that which can harm. Unable to know when she was hurt, Sarah injured herself repeatedly. Whenever she bruised herself playing, her mother had to put a heavy bandage on the bruise so it could heal. When she was about a year old, Sarah had begun biting her teeth out, unable to tell when she chewed something too hard. Without pain to warn her, she could not recognize danger.

Studies suggest that pain may be the most complex of all the senses. Pain detection starts at special receptors lying just below the skin's surface. Injury, extreme pressure, excess heat or cold, and other disagreeable stimuli trigger the receptors to send a nerve signal. The harsher the stimulus, the stronger the signal.

The pain message, like those from all touch receptors, travels to a part of the spinal cord—the dorsal horn—that receives sensory information from throughout the body. The dorsal horn relays it to the brain's thalamus, where distinctions are made between pain and other touch sensations. The thalamus then sends pain signals to the cerebral cortex, the thinking part of the brain. The cortex, interpreting the signal, sends its own reactive signals down the spinal cord and out to the muscles for quick action—for example, the immediate movement of a burned finger away from a hot stove.

But the pain system does not docilely follow this simple model.

Other elements under the brain's control can modify the initial hurt signal at various points en route by releasing natural painkillers known as enkephalins into the dorsal horn. These reduce the intensity of the pain signal or block it completely.

If the signal gets through the dorsal horn, another obstacle awaits it at the base of the brain, where a netlike structure of nerves can squelch incoming transmissions. This pain-suppressing network seems to operate when the injured party concentrates on something besides the pain stimuli—biting a lip, for example, may sometimes suppress pain somewhere else, and soldiers are said to ignore great pain in the distracting heat of battle. □

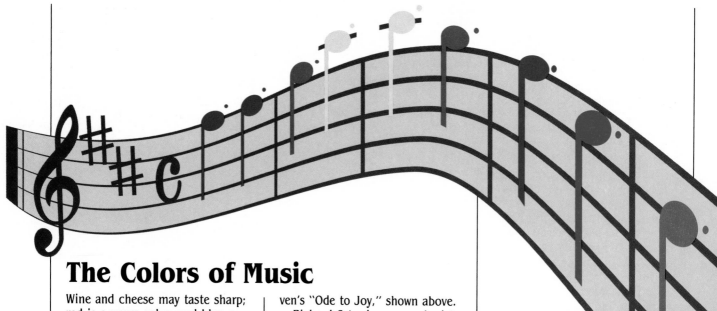

The Colors of Music

Wine and cheese may taste sharp; red is a warm color, and blue a cool one. Trumpet music is bright, jazz is cool, and salsa is red hot. To say, "I see what you're saying," is to understand. For a very few people, however, such metaphors are reality. "What first strikes me is the *color* of someone's voice," one man says, adding that an acquaintance "has a crumbly, yellow voice, like a flame with protruding fibers." Another person sees flavors. "Spearmint tastes like cool, glass columns," she says. "Lemon is a pointed shape, pressed into my face and hands. It's like laying my hands on a bed of nails."

This carnival of sensations marks a rare condition called synesthesia, discovered more than two centuries ago. The senses of synesthetes are joined so that the person actually sees words, tastes colors and shapes, and feels flavors. To a synesthete, the sound of music may be perceived as color, with each note assigned a different hue, as in the opening measures of Beetho-

ven's "Ode to Joy," shown above.

Richard Cytowic, a neurologist in Washington, D.C., has conducted experiments with synesthetes and determined that the linking of the senses occurs because of some unique physical condition in the brains of these people. He has found, for example, that blood flow to some parts of the brain, normally increased by sensory stimuli, decreases in synesthetes during sensory stimulation.

Synesthesia can take many forms. One synesthete, for example, smells colors. "I remember at age two my father was on a ladder painting the left side of the wall. The paint smelled blue, although he was painting white. I remember to this day thinking why the paint was white, when it smelled blue."

To another synesthete, however, colors are associated with sounds. "I am a sight-sound synesthete, most often seeing sound as colors, with a certain sense of almost pressure on exposed skin when sounds are very light or colors very

bright," she wrote Cytowic. "One of the things I love about my husband are the colors of his voice and his laugh. It's a wonderful golden brown, with a flavor of crisp, buttery toast, which sounds very odd, I know, but it is very real."

Synesthesia may seem like a wonderful thing to possess, a kind of omnisensory way of interacting with the world, but it can also be isolating. Some synesthetes worry that they are slipping toward mental illness and suffer the ridicule of friends and family, who often cannot imagine the unique synesthetic experience. "It's like explaining red to a blind person or Middle C to a deaf person," said one synesthete, adding that he found it "a wonderful addition to life and would hate to lose it." □

A Head for Navigation

Tales abound of people possessing an uncanny ability to find their way without use of compass, maps, or any other conventional tool of navigation. Some researchers have tried to explain such talents by looking for a compass, in the form of a small deposit of magnetic mineral, in the human nose.

Other types of organisms have been found to have small quantities of magnetite in what may be a directional organ. Scientists have speculated that such an organ helps bacteria to orient themselves in the earth's magnetic field and may steer birds on their vast migrations.

Explorers to remote places have long reported that the natives can find their way in trackless wastes, a facility that, to the newcomer, must appear to be a magical sense of direction. But these people are in reality just very keen observers of their world.

An example comes from famed explorer F. Spencer Chapman. One day, he recounts, he was kayaking along the east coast of Greenland with a group of Eskimos when a dense fog suddenly developed. He was quite worried that his companions would not be able to find the narrow entrance to their home fjord, but his companions seemed to be unperturbed. Sure enough, after an hour of paddling, the ◊

lead boat turned for shore and before long the party was inside the fjord, safely home.

Chapman soon discovered the Eskimos' secret. "All along this coast," he wrote later, "there were snow-buntings nesting, and each male bird used to proclaim the ownership of his territory by singing his sweet little song from a conspicuous boulder. Now each cock snow-bunting had a slightly different song, and the Eskimos had learnt to recognize each individual songster so that as soon as they picked out the notes of the bird who was nesting on the headland of their home fjord, they knew it was time to turn inshore." □

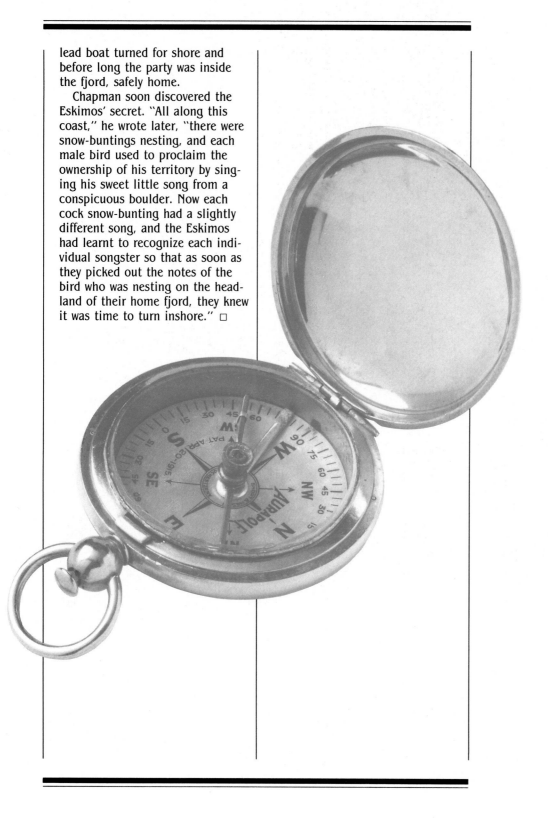

SINGULAR AFFLICTIONS

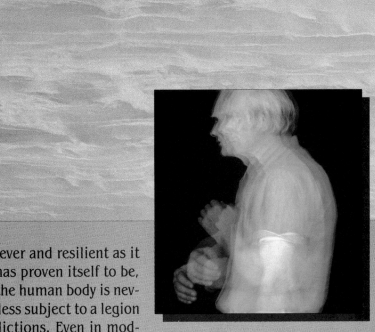

Clever and resilient as it has proven itself to be, the human body is nevertheless subject to a legion of afflictions. Even in modern times, epidemics can sweep through populations like scythes, propelled by bacteria or by the elemental virus, an organism that achieves life only in the cells of hosts, which it cripples or kills. Less dramatic but just as deadly, some illnesses seem to be inevitable, part of the price of living. A medical truism holds that if a person lives long enough, he or she will die of cancer or heart disease. Such voracious consumers of human life are the masters of our mortality.

Also at large are disorders that strike only rarely, many of them notable more for their odd symptoms than for their deadly effect. Bodies may give up their vital spark to the terrors engendered by dreams. They may take to glowing or even, according to some reports, burst into flames. Illness may have its source in how one makes a living or in what one eats or from the kind of pets one keeps. Disease may be manufactured as a weapon or passed down through family bloodlines, hidden like a time bomb in the genes.

Whether epidemic or oddity, the success of any malady depends on the strategy of the bacterium, virus, parasite, or faulty gene that carries it. The invader may kill the infected body so swiftly that it destroys itself as well, or it may inflict just enough weakness and malaise to keep the host going in a debilitated but viable state. It may live quietly within, sharing meals on the sly. Or, as frequently happens, the affliction may lift its siege and vanish, daunted by the body's defenses or, perhaps, tired of its host.

The Global Sickness

Named by Florentine physicians who thought it manifested the influence of planets and stars, influenza has been part of the human scene for twenty-five centuries. Always present to some degree, influenza, or flu, is generally not much more threatening than a bad cold. Even so, only the Black Death—which decimated Europe in the Middle Ages—was more deadly than the great influenza epidemic of 1918 and 1919. Some 20 million people died worldwide, half a million in the United States alone.

The secret of flu's endurance is its versatility. The virus changes its molecular form with each new outbreak. It is never dormant, has no geographic limitation, and quickly sidesteps the immunity conferred by vaccination by changing itself into something else. Flu can penetrate the tightest quarantines and, once contracted, cannot be kept from running its course.

In the 1918-1919 pandemic, the disorder went under various aliases. Soldiers in Flanders called it the Flanders grippe, and it was named Chungking fever where Chinese had worked in France. Japanese sailors called it wrestler's fever; Germans said it was the Blitz katarrh. The name that stuck, however, was perhaps the least threatening: the Spanish influenza. As one soldier wrote home, it sounded as innocuous as a dance.

In the spring of 1919, the Spanish influenza virus vanished as suddenly as it had appeared. Researchers have never been able to identify the lethal strain, and no one knows if it still exists. □

Though ineffective, gauze masks became required apparel during the influenza epidemic of 1918-1919. Here, white-masked police officers go on duty in hard-hit Seattle, Washington.

A Sickly Isle

Anthrax is one of the oldest and most destructive diseases known to humankind. *Bacillus anthracis,* the bacterium that causes the fatal blood infection, can live for many years in marshy soil, resistant to heat, cold, drought, and even disinfectants. Pastured animals such as cattle and sheep are particularly susceptible to anthrax because they eat the grasses growing in contaminated ground. The malady, however, is not just the disease of grazing animals; it spreads readily from an infected carcass or hide to a human body, with the same fatal effect.

Because of its resilience and virulence, anthrax has been one of the biological agents explored for possible use as a deadly weapon. During World War II, the British government began secret experiments with anthrax spores on tiny Gruinard Island, located three miles off the Scottish coast. In 1941, the island's small population was evacuated and replaced with tethered sheep. Anthrax canisters were exploded, and the dying sheep closely observed. Afterward, researchers destroyed the infected carcasses in order to preclude an accidental epidemic.

Unfortunately, anthrax was not through with Gruinard Island. In 1943, the year that the tests were completed, an anthrax epidemic struck the Scottish mainland. It was presumed that spores from Gruinard Island could have washed or blown ashore, and news of the outbreak was suppressed as a matter of security.

Quietly and quickly, experiment-

ers went back to their infected island, and they began burning the heather, hoping thereby to destroy the anthrax once and for all. Subsequent testing revealed, however, that the hardy spores had percolated down into the soil, where they continued to flourish. By then appropriately known as Anthrax Island to the locals, the rocky isle was declared off-limits.

Every few years, scientists re-

turned to Gruinard to test the soil, hoping to reopen the island. As recently as the 1970s, they found three acres still rich in anthrax. Detoxification efforts continued, and in 1987, animals were allowed to return. When none came down with anthrax, the following year—more than forty-five years after the anthrax-weapons experiment—the island was approved for human rehabitation. □

Too Much Too Soon

"In the year of our Lord God 1485," chronicled the sixteenth-century British physician John Caius, a disease was abroad in the land that "immediately kills some upon opening their windows, some while playing with children in their street doors, some in one hour, and many in two." Caius described

the ailment as beginning with a high fever and severe joint pains, "whereupon, by chance, there follows a sweat." Nausea and vomiting followed the sweating, along with rapid heartbeat and pain in the lungs and the head. Victims of the disease seemed mad or at least irrational for a time, then sud- ◊

denly sleepy. The lethargy was the final symptom before death.

Called the sweating sickness, the deadly disease marched into London with Henry VII's troops on August 8, ending the Wars of the Roses with an epidemic that killed thousands in just two months. By October, casualties included London's lord mayor, his successor, and six aldermen. Doctors vainly prescribed tobacco, limes, and purgatives. Patients were routinely bled, which may even have hastened their deaths.

The disease spread faster than undertakers could deal with its victims. Corpses were stacked and buried eight deep. Small towns that lost only half their population were considered to be lucky.

When the English sweat, as it was known on the Continent, crossed into Germany, it killed 15,000 in just five days.

Unlike many illnesses, the English sweat conferred no immunity. People who had it once often contracted it again. The virulent plague returned four times to England but unaccountably vanished after 1551. Its real identity and cause are unknown to this day.

It may have disappeared, some scientists believe, because it was too successful for its own good. Because it killed its hosts so swiftly, the infectious agent could not multiply quickly enough to survive the destruction it had caused. □

Painfully Noted

Acclaimed New York pianist Gary Graffman had spent much of his life perfecting his technique. But hundreds of thousands of hours' practice imparted more than facility: Endlessly repeated movements painfully altered the musculature of his right hand.

Although such injuries are only now being thoroughly researched, early surveys indicate that about three quarters of professional musicians suffer from some job-related affliction. Further, since musicians compensate for impaired performance by practicing even harder, they tend to aggravate the condition. Evidently, the hu-

Art Attack

In Florence, Italy, in the late 1970s, curators of some of the world's greatest art treasures noticed that foreign visitors were fainting in the presence of masterpieces. Because the famous nineteenth-century French novelist Marie-Henri Beyle had himself been emotionally overwhelmed by the beauty of the Santa Croce Church, an Italian psychiatrist gave the disorder Beyle's famous pen name: Stendhal syndrome.

Its victims have fainted in front of Michelangelo's *David* beneath the cupola of the Duomo cathedral, and before Caravaggio's painting of the Roman god Bacchus. The swooners usually are unmarried and unused to traveling. When they recover, they recall sweating, sexual arousement, heart palpitations, and stomach cramps just before passing out. Some sufferers hallucinate voices coming from the paintings.

Dr. Graziella Magherini, who gave the syndrome its name, believes her city's powerful evocation of the past may disorient tourists and induce panic attacks. But skeptics suggest that the unfortunate visitors are being overwhelmed less by Florence than by the crowded, hurried pace of modern travel: What looks like dangerously heightened sensibility may be only jet lag. □

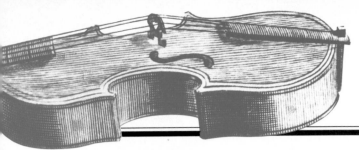

man body was not designed for the extended repetitions and the awkward positions musicians must assume in order to play.

String players, especially violinists, are the most frequent victims. Because the slightest error in their fingering or bowing can push a note off-key or cut it short, they must hold their instruments precisely. The resulting tension produces pains in the upper torso, hands, arms, and neck. Women with long necks and drooping shoulders—the epitome of violinists—are particularly susceptible.

Violins and pianos are not the only hazards. The breathing discipline exacted by oboes and trumpets can induce larynx and blood-oxygen disorders, while the stretched positions required to play the flute can cause rotational problems in the shoulder joint.

For many musicians injured on the job, therapy can undo what music has done, and most benefit from rest without practice. In some cases, the instrument can be modified to eliminate or minimize damage. For example, some violinists use shoulder pads that reduce physical stress.

Unfortunately, the fact remains that many musicians have no recourse but to find ways of living with their impairment. Gary Graffman, for example, now performs piano compositions played with only the left hand. □

A Dermal Palette

In 1960, a Memphis man visited his doctor complaining of chronic stomach pains. While examining the patient for the source of the discomfort, however, the physician could not miss noting another symptom that the man had not even mentioned: His skin was colored a lurid orange.

Diet and disease often alter skin color. A malfunctioning pituitary gland causes pallor, liver damage brings the yellow of jaundice, thyroid dysfunction creates a perfect tan, and congenital heart disease effects a bluish cast. Eating too many carrots can cause a yellowing of the skin called carotenemia, after the yellow pigment carotene, which gives carrots their orange color. But

carotenemia does not go so far as to produce orange skin—nor did anything else that the Memphis doctor could think of.

The physician came to discover, however, that his ruddy patient not only consumed huge quantities of carrots but also great numbers of tomatoes. Tomatoes get their red color from the pigment lycopene, which, like carotene, passes through the body's systems to the skin and attaches to subsurface fat cells. The patient had inadvertently mixed his odd skin color as a painter would have, combining the red hue of lycopene with the yellow of carotene to ultimately produce orange. □

Plastic and Peeling Skin

Predisposed to wrinkle, tan, scar, itch, rip, and sicken, the body's largest organ—the skin—can also be stretched and shed.

In the condition called chalazoderma, skin becomes unusually plastic. When the ailment is congenital, the skin looks normal but can be stretched much like Silly Putty™ before it snaps back into place. Some people with chalazoderma can easily pull the skin from their forehead over their nose like a mask. As an acquired disease, however, the affliction causes the skin to hang from the body in unelastic strands, as though it were melting.

A second skin condition, toxic epidermal necrolysis, causes dangerous blistering. The skin peels off in large sheets, often taking its protective mucous layer with it. The layers of dead outer skin may come away in one piece, from the elbows to the cuticles, like the castoff skin of a snake. The condition's cause is unknown. □

Faded Glory

Henry Moss was a free black living in colonial America. After fighting against the British in the Revolutionary War, he moved to Maryland, married, and farmed. That would have been his quiet destiny had he remained as he was. But, in the early 1790s, Henry Moss began to change color.

First his fingers, then his hands, wrists, and arms faded to pale ◊

whiteness. His abdomen and chest bleached out in large patches. By 1796, Moss had become completely white, his tight black curls a mop of limp pale strands. Sensing fame at hand, Moss quit farming and went to Philadelphia to seek his fortune. There, appearing at the Sign of the Black Horse on Market Street, he became a star.

Along with a public sensation, Moss caused a great medical and philosophical stir. Was his former blackness merely the mark of Africa's hot climate, and did it vanish when one moved to a more temperate world? Would he turn black if he went back? And what about white men living in Africa—would they eventually become black?

In fact, Henry Moss's condition was neither cultural nor geographical. He was afflicted with a rare disease called vitiligo, in which the skin loses the cells that produce the brown pigment melanin. As the cells vanish, white skin patches appear symmetrically on both sides of the body, and the hairs over these areas turn white as well.

Although the cause of vitiligo is unknown, there is evidence that it arises when the immune system malfunctions, causing the thyroid and adrenal glands to produce abnormally large numbers of antibody cells. Aside from their much-altered appearance, however, vitiligo victims are generally healthy. The only known color restorative is cosmetic makeup.

As for Henry Moss, his star faded quickly despite the profound questions that his transformation had raised. He left the city of Philadelphia and headed south, where he supported himself until his death by appearing in the country saloons of rural Georgia. □

Driving King George Crazy

The "madness" of England's King George III began in 1762 when he was twenty-four. A cold and a hacking cough were followed by chest and abdominal pains, rapid pulse, constipation, cramps, muscle stiffness, and a burning sensation throughout his body. Before his doctors could find an explanation for the illness, George recovered. Three years later, a more serious attack disabled him for seven months and kept him from governing.

This time all his earlier symptoms were intensified. Blisters on his legs were so painful he could not bear the touch of clothes or bedsheets. He trembled, hallucinated, and foamed at the mouth. He was unable to sleep or walk and had a terrible sensitivity to bright lights and noise. An incessant talker even when well, the royal patient now rambled on for hours. When doctors inspected his "animal functions," they found his urine was a deep red.

Recovering that summer, the king stayed healthy until 1788, when a five-month-long attack struck him down once more. This time, his doctors decided that George's symptoms must denote insanity and prescribed for their sovereign accordingly. The king was tied to a chair, which he dubbed his coronation chair, or to his bed. He was regularly purged and bled and was kept in dark solitary confinement. The more the

king reacted against this rough treatment, the surer the doctors were of his delusional state. George suffered three more acute attacks before he died in 1820, deaf and blind.

Until recently, history has used the diagnosis of madness to explain George III's inept handling of the American colonies and other incompetencies of his reign. But a recent reexamination of the medical records suggests the monarch may have suffered from a rare, inherited disorder called acute intermittent porphyria. Symptom by symptom, according to his and his physicians' journals, the king's case emerges as a classic, textbook description of a disease once thought to have been unknown before the twentieth century.

Triggered when the blood produces too many of the dark red blood pigments called porphyrins, porphyria causes abdominal cramps and vomiting, burning sensations, convulsions, psychoses, numbness, paralysis, and an intense sensitivity to light. Urine color is the disease's signature. When left in light, a victim's urine turns a deep red. □

Mad Hatters

It was no coincidence that Lewis Carroll made one of the more eccentric characters in *Alice's Adventures in Wonderland* a hatter. In the nineteenth century, craftsmen who made hats had long been known to be excitable and irrational, with a tendency to mix up their words and tremble with palsy—hatter's shakes, the disorder was called. Their trouble was not psychiatric, however, but occupational—a steady poisoning of hatters by their hats. Artisans who daily turned and coaxed felt into fashion were exposed to mercury in the solution used to treat the material. While they worked, the hatters chewed tobacco and spat constantly, in a futile effort to purge themselves of the toxic stuff.

A potent poison to all life forms, mercury becomes airborne at room temperature and can be breathed in through the mouth or nose. Soluble in liquids, it enters the blood and is carried to the liver, brain, and kidneys. Gradual buildup causes chronic mercury poisoning, or mercurialism. Marked by a metallic taste in the mouth, swelling salivary glands, tender gums, loose teeth, and twitches in the eyelids and the tongue, the disorder may lead to muscular tremors in the hands and the feet, headaches, stomach pains, coughing, sweating, and anemia. Attacking the central nervous system, the toxin leads to the behavioral symptoms that gave rise to the expression "mad as a hatter."

The hazards of handling mercury have been known for two thousand years. Conscious of its dangerous effects, the ancient Romans used only convicts to mine it. In 1665, one of history's first labor laws made it illegal for miners to work more than six hours a day in the mercury mines of what is now Yugoslavia. But it was not until the twentieth century that the hat industry eliminated mercury from its processing, thus "curing" the madness of hatters. □

Earthy Fare

In the rural American south, dirt eating—known technically as geophagy—is scorned as a perverse remnant of outworn folkways, a habit practiced only by the ignorant and the poor. Despite the onus, however, geophagy survives—and not without reason. Some kinds of dirt are cheap and effective antacids.

American geophagy has its roots in western Africa, where the pharmaceutical benefits of clay have been known for centuries. Today, the best clay, called *eko* in Nigeria, is mined, cleaned, freed of grit, sun-dried, and sold throughout the continent. Pregnant women use the substance to relieve nausea, while the rest of the population take it for general stomach distress and mild dysentery.

A mixture of quartz and kaolinite, eko absorbs toxins and bacteria from the stomach and the intestines and helps coat the digestive tract with a protective lining. Kaolinite is, in fact, the active ingredient in a popular American commercial preparation for relieving diarrhea.

In the United States, the closest thing to eko is a white clay found in Mississippi. Gourmet geophagists enhance its sour taste by adding vinegar or salt.

It is essential to note, however, that less discriminating dirt eaters are as much at risk as indiscriminate consumers of mushrooms. People who eat the wrong kind of dirt may suffer severe anemia or even death. These hazards exist when the soil's acidity and composition impede the ability of the body to digest such essential minerals as zinc and iron. □

An Incendiary Enigma

By 9 p.m. on July 1, 1951, Mary Hardy Reeser had already taken her sleeping pills and dressed for bed, chatting with her physician son and his wife and with Pansy Carpenter, landlady of her St. Petersburg, Florida, apartment. When her visitors left, the plump sixty-seven-year-old was resting in an armchair facing an open window. She wore a rayon nightgown, cotton housecoat, and satin bedroom slippers. She was smoking a cigarette.

The next morning, after a Western Union messenger failed to rouse Mary Reeser, Mrs. Carpenter went by for their customary morning coffee. She knocked and got no response. When she tried to enter, she found the doorknob too hot to touch. Alarmed, she got two house painters to force their way into the apartment.

There, on the spot where hours before had been a woman, a chair, and an end table, they found only some heat-eroded coil springs, a shrunken skull, a charred liver, a piece of spine, and a foot in a slipper, all in a pile of greasy ashes. A 175-pound woman had apparently been reduced to less than ten pounds of charred remains. Yet, though the body had been almost completely incinerated, the underlying carpet was only singed and little else in the room was burned. Starting about four feet above the floor, an oily soot covered the curtains, walls, and ceiling. A mirror had cracked, presumably from intense heat, but newspapers and linens near Mrs. Reeser's remains seemed to be untouched.

The puzzling case of Mary Reeser is the best-

Although little else was damaged in the room, a slippered foot and ashes are all that remained of Pennsylvania physician John Bentley, who mysteriously caught fire on December 5, 1966.

documented among more than a hundred reported incidents of bodies inexplicably consumed by fire in a phenomenon called spontaneous human combustion. People are alleged to have burst into flames while afoot or in cars, in cribs, beds, and armchairs, even on a London dance floor. The most famous case, however, was the purely fictional death of a boozy rag-and-bottle seller named Krook in Charles Dickens's *Bleak House*—a demise described in excruciating detail by Dickens and strikingly echoed by the Reeser case.

The victims of spontaneous human combustion are usually overweight, elderly, and inclined to drink to excess. The fires alleged to consume them are described as flickering, pale blue flames that destroy only the torso and head, leaving the limbs uncharred—just the opposite of what happens in a normal burning. The conflagration, which must reach several thousands of degrees to reduce a human body to ash, typically causes negligible radiation or combustion damage to objects nearby. In the nineteenth century, it was theorized that the mysterious flames erupted out of alcohol and body fat, burning as a kind of candle, or

from flammable gases generated in the intestines. But experimental efforts to ignite such fires have gone against the idea of spontaneous combustion: The body, mostly salt water, is very hard to burn.

Other students of the phenomenon argue that static electricity could trigger human combustion. About one person in one hundred thousand can build up a charge of as much as 30,000 volts, more than enough to spark an explosion in a room full of volatile gases. However, there is no evidence that such detonations could incinerate a body while leaving the contents of a room untouched.

Alternatively, vitamin B10, or inositol, a compound about as stable as nitroglycerin, has been called a possible trigger. The body makes this phosphate-potassium compound as a supplementary energy supply to be tapped if it runs short of glucose. Sedentary people, as the elderly and overweight are likely to be, may produce too much inositol, building up a dangerously high concentration.

However, inositol was discounted as a factor in the Reeser case by Wilton Krogman, a well-known forensic anthropologist brought in by the FBI. As he had on other

occasions, Krogman tried to reproduce the cause of death in his Pennsylvania laboratory, burning dozens of cadavers in hickory wood, coal, gasoline, and kerosene fires. None of these ignited the body, nor would a standard cremation, which applies temperatures of 2000 degrees Fahrenheit for eight hours. Only in a pressurized crematorium furnace capable of heating a body for eight hours at over 3000 degrees Fahrenheit do human bones finally turn to ash.

Krogman sneered at spontaneous human combustion as a cause of death. Instead, he believed the woman had been taken somewhere and cremated at very high temperatures, then restored to her room, where the killer added such touches as the oily soot and singed carpet. In Krogman's view, the murderer had to be very knowledgeable about burning things—and perhaps a Charles Dickens fan.

The Reeser case has never been solved, but skeptics scoff at attributing the death to a cause as improbable as spontaneous human combustion. There is no great mystery, they contend, in the death by fire of a sedated woman, wearing flammable nightclothes, last seen smoking a cigarette. □

The Glowing Woman of Pirano

In early 1934, Dr. Giocondo Protti was summoned from Venice to observe a forty-two-year-old woman convalescing at the Pia Casa di Ricoveri in the nearby Italian coastal town of Pirano. The convalescent, Anna Monaro, was not recovering from any particular illness

but was allegedly suffering from a most peculiar symptom: Her body emitted colored lights.

The report from Pirano was one of hundreds of eyewitness accounts of glowing people that have come down through the centuries. Saints, spiritual mediums, and the

dying, in particular, have been said to luminesce. Sathya Sai Baba, a twentieth-century Indian mystic, was seen by thousands of followers to have bright red rays emanating from his head, and in the 1940s, the Russian mystic Seraphim of Sarov reportedly shone with "a blinding light spreading several yards around." Soft, phospho- ◊

rescent emanations were a fairly commonplace, if inexplicable, phenomenon during the spiritualist séances that were popular in the late nineteenth and the early twentieth centuries.

The case of Anna Monaro was different, however. Because Protti and his colleagues at the hospital were able to study her in a controlled, scientific setting, Monaro's is the best-documented example of human luminosity thus far.

She began to emit light during a strict Lenten fast, usually in a restless interval of sleep, and always unconsciously. The emissions ranged from red to green and varied in intensity. Never visible for more than a few seconds at a time, the light changed shape, sometimes becoming a fan, sometimes a glowing free form, always originating near her chest, and often visible from the hallway some thirty feet away from her bed. After each occurrence, the woman would moan and awaken as if from a nightmare, invoking Jesus' help.

Protti set up equipment to measure possible electromagnetic fields during the emissions and attempted to photograph the lights. No trace of electricity was obtained, but Protti's strip of exposed film showed a cloudy darkening that corresponded to Anna Monaro's luminescence. Other tests showed that, except for high blood pressure and a high level of radiation in her blood—a finding later called into question—she seemed to have no physical problems.

Protti was unable to explain the phenomenon he had observed, nor could scientists with Italy's National Research Council when they examined the glowing woman a few weeks later. Monaro apparently did not understand her remarkable affliction either, but she did seem to have some feel for its behavior. According to Protti, the last time he saw her she made a prediction. "Lent is over and I feel less heavy," she said. "I feel that the light will never come back again." From all accounts, it never did. □

Witch's Milk

Milk production, or lactation, normally occurs in new mothers. But it can also take place in women who have no obvious reason to lactate, in men, and even in babies. Lactation by infants was so shocking that its product was once referred to as witch's milk, suggesting dark magic behind its improbable appearance.

The medical term for inappropriate milk production is galactorrhea. The condition happens when the pituitary gland, the master switch for the hormone system, orders too much prolactin, the hormone that tells the body to start making milk. This usually happens when drugs miscue the pituitary gland. For example, estrogen, a hormone that figures in pregnancy and other female functions, is present in oral contraceptives and can send a false message to the pituitary gland to produce milk. Some agents in tranquilizers stimulate a similar response, regardless of sex. Milk production may also be prompted by a pituitary tumor or diseased thyroid. □

The Mulford Inheritance

In 1630, three men who had been persecuted as witches in Britain arrived in the American colonies. But unlike their fellow immigrants, they found little peace in the suspicious world of seventeenth-century New England. The taint of witchcraft persisted, and several of their family members were executed. Not for more than two centuries did the family learn that its alleged evil was the manifestation of an inherited disease.

One of the original trio settled in East Hampton on New York's Long Island. His descendants, the Mulfords, became known locally not for evil deeds, but for tragic affliction. To young George Huntington, however, they were a fascinating case history. As a boy, he had visited the Mulfords with his father and grandfather, both general practitioners in East Hampton, and he kept his interest in the family through his own graduation from medical school. In 1872, newly a doctor, Huntington published a paper describing the Mulfords' complaint, a terrible illness since called Huntington's chorea.

Huntington had noted that the family suffered from a disorder that always struck sometime between the ages of thirty and fifty and seemed to be inherited. The disease progressively destroyed the victim's motor skills, intelligence, and personality, and was ultimately fatal. Huntington labeled it chorea, a term describing any disorder that causes the body to jerk and dance involuntarily.

But Huntington never comprehended the full horror of the illness that bore his name. Because it is carried by a single gene, each descendant in an afflicted family has a fifty-fifty chance of contracting it. The disorder does its work by killing vast numbers of cells near the center of the brain, destroying, in ways that are still imperfectly understood, movement, thought, and memory.

Huntington's chorea is extremely rare, with only 4 people in 100,000 affected in the United States. It can be contracted only through genetic transmission and begins insidiously with mild depression and irritability. As the disease evolves, its victim's behavior comes to resemble that of a drunk, with slurred speech and swaying stride. Eventually, involuntary convulsions and swallowing difficulties confine the sufferer to a wheelchair, and depression deepens to despair. Among known and prospective victims of Huntington's chorea, the suicide rate is 200 times the general population's.

Until recently, the only way for children to know whether they had inherited the Huntington gene was to wait for its symptoms to appear. For some, this meant waiting a lifetime, afraid to have children. Now, a chromosomal test developed by Johns Hopkins University and Massachusetts General Hospital can detect the gene. Possible heirs to the disease can learn the truth while they are young—although some prefer not to know.

All the news from the tests is not bad. If Huntington's chorea misses a generation, the gene is gone forever. □

Sleep Attacks

Narcolepsy, named in 1880 by its French discoverer, Jean Baptiste Edouard Gelineau, means literally to be overtaken by sleep. It describes a rare illness in which the afflicted are prone to sudden, irresistible urges to nap.

The sleep attacks can come at any time, catching the victim in midstride—at the dinner table, walking down the street, or, dangerously, at the controls of a moving vehicle. The narco-leptic has several brief episodes a day. The attacks usually last from ten to thirty minutes. On waking, the afflicted person feels rested at first but soon succumbs to the persistent fatigue of narcolepsy.

The attacks are often accompanied by emotional surges, which may cause uncontrollable mirth or depression, and cataplexy, an abrupt loss of muscle tone. Sufferers may also experience paralysis and hallucinations. Of unknown cause, narcolepsy usually begins in adolescence and lasts a lifetime. □

Narcoleptic Ray Johanson dozes during a sleep attack at the National Narcolepsy Association's headquarters in California.

Touretters

In 1885, French neurologist Georges Gilles de la Tourette described a malady that continues to astonish a century later—Gilles de la Tourette's syndrome, a disorder marked by uncontrollable excesses of nervous energy, extravagant motions, and all manner of physical and vocal tics. Touretters, as victims of the disease call themselves, endure lives that, on the one hand, seem to transcend normal life in animated intensity. On the other hand, their behavior threatens at any moment to fly completely beyond their control.

The ailment usually begins between the ages of two and fifteen, often showing itself at first as an exaggerated blink or slight facial tic. Once under way, however, Tourette's syndrome quickly swarms the patient with multiple, involuntary tics that twitch the face and limbs, spreading from the head downward through the body.

At the same time, the verbal equivalent of muscular tics appear as sniffings, grunts, and explosive barks. Many experience what is called coprohenomena, the impulse to utter or write obscene words, or make obscene gestures. The afflicted may also mimic the words and motions of others, repeat phrases rapidly, touch things, become obsessed with numbers and counting, and, more seriously, engage in self-mutilation.

The illness is incurable and in most cases lasts a lifetime, although it may alter its form. For example, one kind of tic may disappear, only to be replaced by another. Touretters are also able to control the disorder by an act of will, but only for a while. Suppressed tics build up like steam in a boiler; when the victim relaxes, the bottled-up excitement explodes into view.

In the Middle Ages, Touretters were thought to be possessed by the devil. One fifteenth-century priest with the illness compulsively thrust out his tongue. He was reportedly cured by exorcism. Most victims, less fortunate than the priest, were imprisoned, chained, and shunned by doctors and society. Despite the fact that Tourette's syndrome was identified as a disease in the late nineteenth century, its cause remained a mystery. Psychoanalysts, Sigmund Freud among them, decided it must be a form of hysteria arising from sexual repression.

In 1961, however, a French physician, M. J. N. Seignot, reported a case in which Tourette's symptoms were apparently controlled by doses of haloperidol, a drug used to treat schizophrenia. Further tests showed almost a 90 percent success rate in controlling Touretters' tics. The illness, researchers found, was not psychological but neurochemical, triggered by an excess of (or an excessive sensitivity to) the neurotransmitter dopamine, a chemical produced in the brain. There is also some evidence that the illness is transmitted genetically; an extended Mennonite family in La Crete, Alberta, has included Touretters for six generations.

But a cure is not entirely welcomed by all of those who are afflicted. Taking a drug may eliminate their bizarre tics, but it dulls the sharp edge of their excited lives as well. One patient solved the dilemma by taking haloperidol during the week in order to work and live normally, then, on weekends, abandoning himself to his symptoms, regaining the sharp repartee and dash of the Touretter. "Suppose you could take away the tics," he asked his physician, "what would be left? I consist of tics—there'd be nothing left." □

Canadian Touretter David Janzen's right arm suddenly convulses while driving, but eight-year-old Barbara, accustomed to her father's uncontrollable but generally harmless tics, hardly notices.

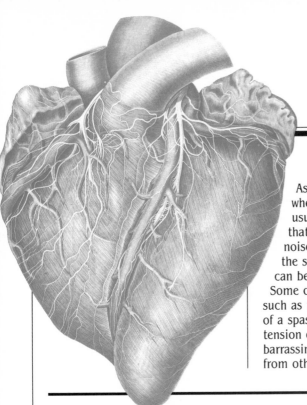

Not-So-Funny Noises

Aside from the occasional wheeze or gurgle, the body usually functions so quietly that we assume it makes no noise. In rare cases, however, the sounds of working organs can be heard across a room.

Some of this unexpected racket, such as nostril clicking—the result of a spasmodic tic stemming from tension or fatigue—is merely embarrassing. But peculiar noises from other parts of the body may be a signal of deeper trouble.

Loud clicks emanating from the chest, for example, may indicate a malfunctioning valve somewhere in the circulatory system. These sounds are called systolic knock. And when the ordinarily silent brain begins making the hushed but audible bellowslike sound known as intracranial bruit, it may be attempting to tell its owner about a blocked or ruptured blood vessel or a tumor. □

Repugnant Residents

Although people of temperate lands may look with horror at bodily parasites that plague inhabitants of the tropics, the truth is that tiny, unwelcome intruders exist at all latitudes. Even in the technically advanced nations of the north, few people go through life without some kind of parasite—a tiny wingless insect under the skin or sundry arthropods scattered through the body's interior. Among the biological hangers-on apt to produce the greatest shudders are the helminths—worms. Not that these repulsive roomers are lethal or even horrifically damaging; they are not. In most cases, the worst part of hosting them rests with knowing they are there.

The common tapeworm, *Taenia saginata (right, life-size)*, is a case in point. Passed to humans from contaminated beef, this relatively benign boarder does not starve or emaciate the host, as legend has it, or grow to be hundreds of yards long. Tapeworms may finally become almost as long as their intestinal home—at least one has been measured at twenty-four feet—and are presumed to live as long as their host unless ejected medically. But the only sign of *T. saginata's* presence is an occasional rectal itch or the discovery of a discarded segment. Nausea and weight loss, if any, result from the victim's revulsion, not from the worm. *Taenia solium*, a very rare tapeworm found in pork, is likewise innocuous when it lives in the digestive tract. However, unlike *T. saginata*, it can move. When it travels to the brain, it can produce symptoms that mimic epilepsy.

Worms generally arrive in food that has not been frozen cold enough or cooked enough to kill them. Any uncooked meat or fish offers a risk of infestation. According to public health researchers, a growing taste for the raw seafood dishes of Japan, Latin America, and Scandinavia has begun to spread flukes and tapeworms found in fish. But these are usually coughed up or vomited up soon after being eaten, causing more dismay than physical distress. □

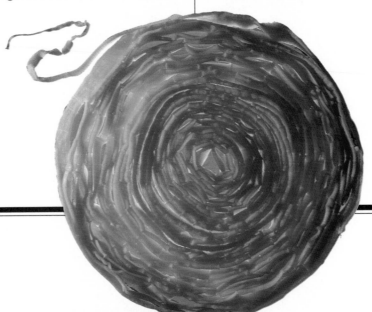

Sloshing Sickness

The pleura is a two-layered membrane that wraps around both lungs and lines the chest cavity. Ordinarily, it maintains a seal that lets the lungs breathe while keeping out infection. Sometimes, however, the seal is pierced, causing breathing difficulties and strange sounds.

One such disorder, called hydropneumothorax, is caused by pockets of water accumulating in the pleura to the extent that the sloshing liquid can be heard some distance away. One London physician related the story of a British huntsman who, in spite of his hydropneumothoracic condition, insisted on riding to hounds. When the man's horse jumped an obstacle, his comrades heard a great splash—not the sound of the unseated rider landing in a brook, but the galloping tides of his waterlogged pleura. □

The Coma and the Cancer

For seven years, the man known as Uncle Toby Oakins sat in a corner of his London home, cared for by his family as though he were a plant or a piece of furniture. Waxlike, cold to the touch, and silent, he did indeed more resemble a fixture than a living being. When a doctor visiting another family member accidentally discovered Oakins in 1957, relatives explained that immobile Uncle Toby had run down like an unwound clock back in 1950. Incredibly, they had simply accepted his diminishing life signs and given him what care they could. When the doctor examined Oakins, he found that the old man was sustained by a slow, almost imperceptible pulse.

Removed to a London hospital, Oakins showed a body temperature of only sixty-eight degrees Fahrenheit, some thirty degrees below normal. Doctors observed that his thyroid gland had virtually shut down. Ordinarily, the thyroid stimulates the body's metabolism, the process in which energy is extracted from food. With virtually no thyroid function, the man's internal fires had almost flickered out, leaving him suspended in a hypothyroid coma called myxedema.

The medical solution to his problem was a straightforward one—dosing Uncle Toby with a thyroid hormone in order to reignite his metabolic furnace and bring him back to active life. Concerned that restoring him too suddenly would damage organs accustomed to little or no activity, doctors administered the cure very gradually, raising his temperature a few degrees at a time.

After a month of this, when the thermometer had passed the eighty-degree mark, Oakins began to move and speak—but not, with any accuracy, to remember. As far as he was concerned, he had been sick a day or two before—back in 1950. Afraid of what effect the truth would have on him, the hospital staff carefully shielded Oakins from the shattering fact that he had lost seven years of his life.

Six weeks later, however, he began to cough up blood. Doctors found a large tumor in his chest, a malignant cancer that typically kills very swiftly. Ironically, the coma that robbed from Oakins's life had also preserved it. X-rays from 1950 showed the unmistakable beginnings of the fast-growing, fatal malignancy, a tumor lulled into dormancy by the same coma that immobilized its host. No longer suppressed by Oakins's inner chill, the reawakened cancer killed him within a few days. □

Victims of Frey's, or auriculotemporal, syndrome perspire when they chew such tasty foods as chocolate, cheese, and heavily spiced dishes. This odd and inappropriate reaction apparently results from damage to the nerve endings of the parotid gland, which controls salivation.

One theory is that the injured nerve endings heal but lose their protective coverings; unshielded, they begin receiving impulses from other nerves. A signal to salivate is construed as the order to perspire.

Patient Zero

In 1984, thirty-two-year-old Gaetan Dugas died in Quebec City, Canada, of acquired immune deficiency syndrome, or AIDS. Dugas's death was not widely remarked, for, by that time, the virus that causes AIDS had been identified and the lethal, incurable disease had been recognized as a potential global epidemic. Like many others, the blond flight attendant had contracted the disease as a sexually active homosexual, and, like many victims, he had helped spread it. Nevertheless, Gaetan Dugas acquired a peculiar celebrity. Two years before his death, the National Centers for Disease Control in Atlanta, Georgia, identified him as the person responsible for bringing AIDS to the United States. Dugas was Patient Zero.

As a steward for Air Canada, Dugas had been an ideally mobile carrier—and a fatally prolific one. He bragged about 2,500 sexual conquests, mainly men in California and New York. When National Centers for Disease Control sleuths began tracing the infection back to its origins in America, they discovered that Dugas had been the sexual partner of the first nineteen AIDS cases in Los Angeles, the first twenty-two in New York City, and eight more in other towns. Warned in 1982 about the threat that his personal lifestyle was posing to other people, Dugas deliberately continued to spread the fatal illness until his death.

Researchers note, however, that AIDS would not have remained trapped in Africa, where it is believed to have arisen among monkeys. Promiscuous sex, virus-tainted needles, and worldwide traffic in contaminated blood products made it inevitable that AIDS would eventually come to the industrialized world. If there had been no Gaetan Dugas, somebody else would certainly have become Patient Zero. □

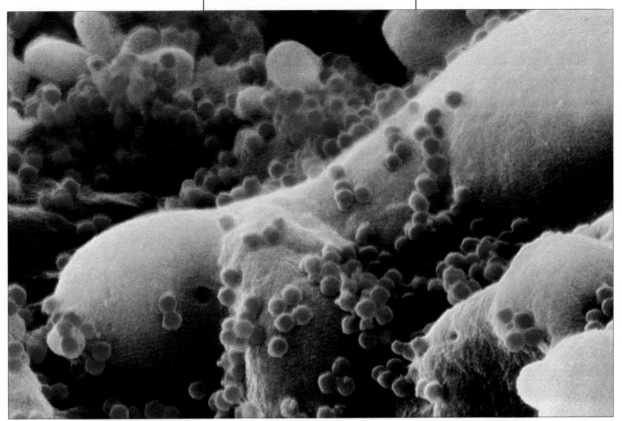

In a computer-tinted image by Swedish photographer Lennart Nilsson, AIDS-bearing HIV viruses (small blue hexagons) swarm across the surface of a T-lymphocyte (yellow), a first-line defender in the body's immune system.

Giving Sickness a Name

To be studied and treated, afflictions must be named, and there is no real system to it. The names of illnesses may spring from popular descriptions, as in smallpox and chickenpox, or the harlequin reaction, which turns the body half red, half pale white. Occupations are likewise identities for disease, from tennis elbow to miner's phthisis and weaver's bottom.

Disease nomenclature is often assembled out of an inventory of standard Greek and Latin parts. For example, the heart paralysis called cardioplegia comes from the Greek *kardia,* for "heart," and *plege,* for "stroke." *Lymphoma,* a general term for diseases of the lymph tissues, comes from the Latin *lympha,* meaning "clear water." Sometimes the parts get jumbled. *Lymphocythemia,* for instance, joins the Latin *lympha* with the Greek *kytos,* for "hollow cell."

Many illnesses are named for the person or persons who first described them—the deadly Waterhouse-Friderichsen syndrome, for example—or for people who suffered them, as when sarcoidosis, a cell disorder, was dubbed Mrs. Mortimer's malady. Sometimes, the immortality is shared. Weber-Christian disease, an affliction of feverish nodules just below the skin, bears the name of the doctor who discovered it, Weber, and that of Henry Christian, who had it.

Maladies such as Spanish influenza, English sweat, Rocky Mountain spotted fever, and Alpine scurvy reflect the geography of their occurrence, while others spring from chauvinism: Syphilis was called the French disease in Britain, the English or Spanish disease on the Continent. Disorders may receive onomatopoeic titles—such as hiccups and belches, named for the sounds the body makes. And appearances are important. Thief of Baghdad syndrome refers to a turban-shaped tumor. □

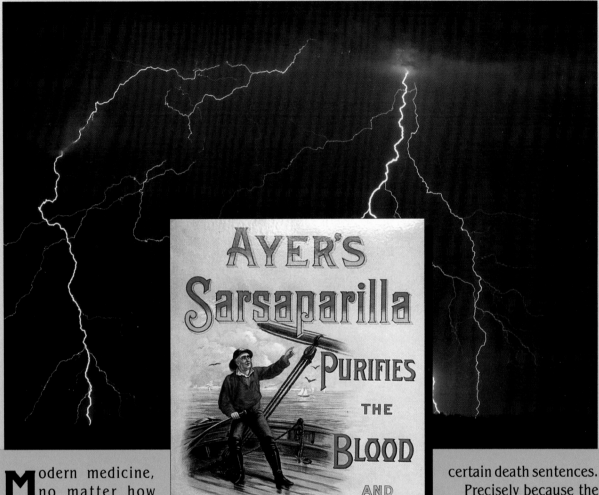

Modern medicine, no matter how far advanced the state of its science, contains elements of mystery. Despite a considerable knowledge of how the machinery of healing operates, physicians are frequently surprised and even baffled by the human body's own powerful impulse to be well and whole. The history of medicine blends news of such scientific breakthroughs as Sir Alexander Fleming's discovery of penicillin with well-documented reports of inexplicable cures and recoveries from what appeared to be certain death sentences.

Precisely because the curing process is but imperfectly understood, it has been a fertile ground for the fanatics and the frauds. A single step takes one across the line to treatments that are derived more from ideology or entrepreneurial instinct than from science. Every new medical discovery has its false echo; every disease that is pronounced incurable inspires the invention of bogus cures. When the body will not heal, there is money to be made from selling hope.

Cocaine—the First Epidemic

Now deemed a highly addictive and dangerous substance, cocaine was once considered a virtual wonder drug. Back in 1884, Sigmund Freud, the father of psychotherapy, dosed himself with coca, the Andean leaf from which crystalline cocaine is extracted, and lauded the effects of the "magical substance" in an essay that became famous. It "wards off hunger, sleep, and fatigue," he wrote, "and steels one to intellectual effort . . . [producing] no compulsive desire to use the stimulant further." Freud was particularly intrigued by the possibility of introducing opium addicts to cocaine to wean them from their habit.

Cocaine entered widespread medical use as an anesthetic. But this potent drug also spread through society, only gradually revealing its darker side. American grocers, saloon keepers, and druggists were offering cigarettes, cordials, elixirs, and other items laced with cocaine. Austrian chemist Albert Niemann had perfected the extraction process in 1860, and by 1885, Parke, Davis and Company, the leading U.S. manufacturer, was marketing the drug in fifteen forms, many purely recreational.

Consumers with expensive tastes could also indulge in imported coca liqueurs such as Vin Mariani, a wine-based French drink that won testimonials from such notables as Pope Leo XIII and the inventor Thomas Edison. Less well-to-do fanciers imbibed a new soda-water concoction called Coca-Cola, with a cocaine content estimated at .0025 percent or greater. They could also pick from a variety of imitators, including Koca Nola, Nerv Ola, Wise Ola, and the bluntly named Dope.

Anticipating Freud, Detroit's *Therapeutic Gazette* in 1880 encouraged the use of cocaine as a remedy for not only morphine addiction but also simple cases of the blues. Other medical organizations followed suit. Because the drug could shrink inflamed nasal passages and sinuses, the American Hay Fever Association gave it highest marks. Patent asthma powders and toothache drops containing cocaine were favored over-

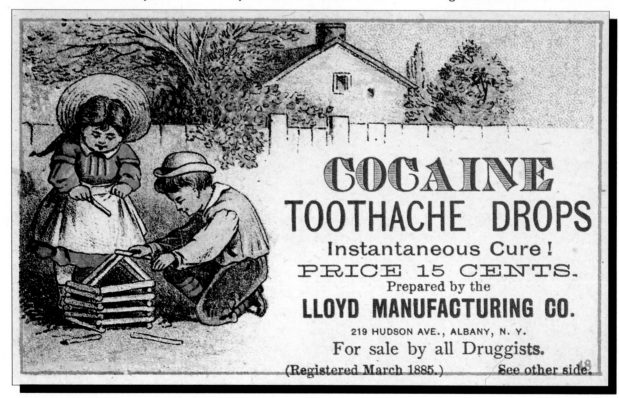

COCAINE TOOTHACHE DROPS
Instantaneous Cure!
PRICE 15 CENTS.
Prepared by the
LLOYD MANUFACTURING CO.
219 HUDSON AVE., ALBANY, N. Y.
For sale by all Druggists.
(Registered March 1885.)   See other side.

the-counter medicines, as common as aspirin is today.

In time, reports of cocaine abuse began to trickle in, but they were commonly viewed with skepticism. For example, William Hammond, a former U.S. Army surgeon general and advocate of the drug, insisted in 1887 that cocaine use "was a case of preference, and not irresistible habit." Other investigators soon proved Hammond to be wrong. Leading the way toward the eventual restriction of the drug, the Connecticut State Medical Society in 1896 documented cocaine's addictive dangers and recommended that it be made available only to physicians as a local anesthetic.

A flood of newspaper and magazine articles, and popular accounts such as Annie Meyers's book *Eight Years in Cocaine Hell,* chronicled the drug's destructive powers. By 1910, the State Department had begun linking cocaine with crime, in addition to fueling unfounded racist fears that black "cocaine fiends" would run rampant over the South. Congress responded four years later with the Harrison Act, which brought the sale and distribution of cocaine and other drugs under federal control. □

POPULAR FRENCH TONIC WINE

Fortifies and Refreshes Body & Brain
Restores Health and Vitality

A Healthy Appetite

During Napoleon's foray into Egypt in the late 1790s, French military surgeon D. J. Larrey noted that the maggots—larvae of flies—infesting soldiers' wounds ate only rotting flesh, leaving healthy tissues untouched. This observation led to the development in the 1900s of maggot therapy.

Just as the ancient Maya and the Australian Aborigines had reportedly done, doctors applied fly larvae to infected wounds to cleanse them and speed healing. Practiced until the advent of antibiotics, maggot therapy has occasionally resurfaced. In 1989, doctors at Children's Hospital in Washington, D.C., used it on seventeen-year-old Shannon Dillingham after antibiotics failed; they saved her life by dressing her gravely infected legs with 1,500 fly larvae. □

Antibiotics in the Backyard

In the early days of surgery, infections so invariably followed operations that doctors thought them part of the natural healing process. Yet, in roughly one in five patients, supposedly harmless infection flared into deadly "hospitalism"— a combination of gangrene, blood poisoning, and a chronic skin disease called erysipelas.

During the American Civil War, a few surgeons seized on the notion that uncleanliness led to hospitalism. They began to combat contamination by using iodine and bromine to disinfect wounds, keeping bed linens clean, and isolating infected patients. Their techniques were not widely imitated, however, and the value of a germ-free surgical environment was not codified until 1867, when a British surgeon by the name of Joseph

Lister set down the principles of medical hygiene.

But in the wake of this new antiseptic philosophy came the confounding claims of an American surgeon, Addinell Hewson, who said he had eliminated infections by packing wounds with dirt. As detailed in his 1872 volume entitled *Earth as a Topical Application in Surgery,* ninety-three patients with gunshot wounds, ulcers, burns, and various surgical traumas received this treatment with complete success.

Although the profession dismissed Hewson's cure at the time, it has recently received scrutiny from medical historians. In all likelihood, they report, the efficacy of the backyard medicine was due to soil-dwelling organisms, such as the bacterium in bacitracin, later tagged as antibacterial agents. Hewson had stumbled upon antibiotic medicine more than half a century before its time. □

The Medicine Is the Medium

Biochemistry was an art form to Scotsman Alexander Fleming, who in 1928 discovered the antibacterial powers of a lowly mold, *Penicillium notatum.* Relaxing in his laboratory at London's Saint Mary's Hospital, Fleming would draw a design on a sheet of blotting paper and coat it with a gelatinous mixture of nutrients called agar. He then went over his outlines with a stylus, creating grooves in the agar that he filled with penicillin, which bacteria could not cross. Setting his living "paints"—the bright yellow *Staphylococcus,* blue *Bacillus violaceus,* and red *Bacillus prodigiosus*—in the nutrient medium, Fleming allowed them to grow within their antibiotic outlines to form landscapes, buildings, and human figures *(above),* and one time, in honor of a visiting Queen Mary, the Union Jack. □

Animal Magnetism

Like many eighteenth-century physicians, Franz Anton Mesmer took his science with a pinch of the supernatural. As a student at Vienna's vaunted medical school, he wrote his doctoral thesis on "the influence of the stars on human constitutions."

Mesmer's astrological interests soon gave way to a fascination with the experiments of the Austrian royal astronomer and Jesuit priest Maximillian Hell, who investigated the effects of magnets on the sick. Mesmer began passing magnets over afflicted bodies to activate the universal healing fluid, or animal magnetism, he believed flowed through all living things.

Soon after moving to Paris in 1779, Mesmer attracted a throng of wealthy devotees, most notably Queen Marie Antoinette, the duc de Bourbon, and the marquis de Lafayette. Dressed in a lilac cape and waving a "magic" wand, Mesmer administered to his aristocratic followers, who sat around a large tub, or *baquet,* grasping protruding rods. This battery was supposedly the source of healing current. As a pianist played in the corner, Mesmer would touch those in the circle with his wand, causing them to laugh or cry hysterically, tremble, or lapse into convulsion-like "crises."

As hundreds proclaimed themselves cured by Mesmer's methods, King Louis XVI ordered the French Academy of Sciences to investigate. A blue-ribbon panel led by Benjamin Franklin, at that time the American ambassador to France, denounced animal magne-

tism as pure imagination. The verdict had little effect on Mesmer's popularity, but the French Revolution put an end to his fashionable séances. Discredited, Mesmer fled to England, thence to Vienna and his birthplace near Lake Constance, where he died in 1815 at the age of eighty-one.

Mesmer's magnetic theories were quickly forgotten, but his powerful, spellbinding personality was not. His surname became a synonym for what this physician did best: In modern usage, to mesmerize is not to heal but to hypnotize. □

Electric Tweezers

In 1796, the United States government granted its first medical patent to Elisha Perkins, a prominent Connecticut surgeon. Perkins's invention consisted of two three-inch rods, one alloyed of gold, copper, and zinc, the other of silver, iron, and platinum *(below, right)*. When passed over the body, Perkins claimed, these "metallic tractors" drew off an electrical "fluid" that caused epilepsy, gout, and rheumatism.

Sold in red morocco cases to such eminent figures as the chief justice of the Supreme Court, the tractors became faddish. Faith in Perkins's medical talents dwindled, however, after he attempted to quash a yellow fever epidemic in New York City in 1799 with a home-brewed remedy and died from the disease himself.

In England, meanwhile, Perkins's son Benjamin used odes, ballads, and cartoons to promote tractor sales. An easy target for satirists *(far left)*, Benjamin was eventually driven from the country—but not before he had extracted a fortune from a number of well-heeled believers. □

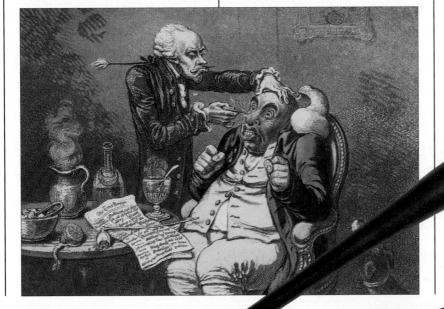

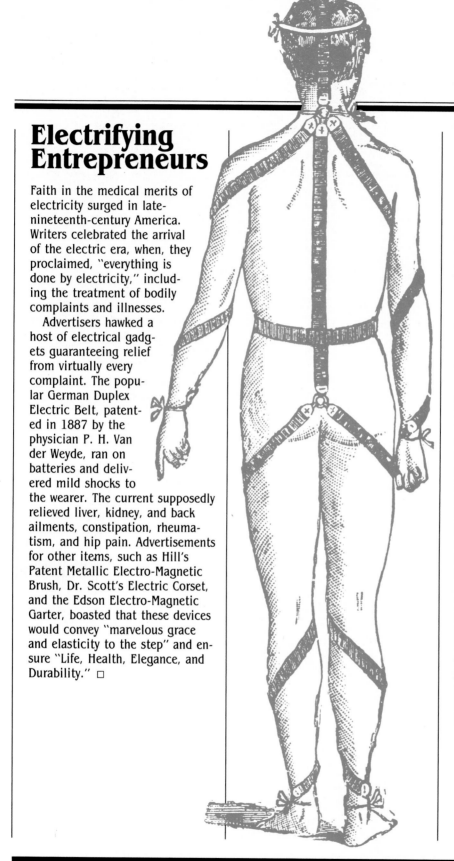

Electrifying Entrepreneurs

Faith in the medical merits of electricity surged in late-nineteenth-century America. Writers celebrated the arrival of the electric era, when, they proclaimed, "everything is done by electricity," including the treatment of bodily complaints and illnesses.

Advertisers hawked a host of electrical gadgets guaranteeing relief from virtually every complaint. The popular German Duplex Electric Belt, patented in 1887 by the physician P. H. Van der Weyde, ran on batteries and delivered mild shocks to the wearer. The current supposedly relieved liver, kidney, and back ailments, constipation, rheumatism, and hip pain. Advertisements for other items, such as Hill's Patent Metallic Electro-Magnetic Brush, Dr. Scott's Electric Corset, and the Edson Electro-Magnetic Garter, boasted that these devices would convey "marvelous grace and elasticity to the step" and ensure "Life, Health, Elegance, and Durability." □

The Black Box

In the 1930s, Hollywood chiropractor Ruth B. Drown began claiming that she could heal without surgery using what she called a "radio therapeutic instrument." Drown would place a piece of blotter paper bearing a drop of a patient's blood into the device, a black box with dials on it, and hook her patient to the machine with electrodes. She then twiddled dials in order to expose the patient to curing "radio waves."

In 1948, Marguerite Rice of Illinois, who had discovered a small lump in her right breast, began a series of treatments for what Drown diagnosed as a fungus in the digestive system. When Rice's condition showed no improvement after some eighteen months of continuous electrotherapy, her husband alerted the U.S. Food and Drug Administration. Tried for fraud in 1951, Drown was banned from interstate commerce. She unsuccessfully appealed the ruling in 1952; Marguerite Rice died of cancer during the appeal.

Prohibited from dispensing any "cures" across state lines, Drown resumed her California practice, and she continued her treatments for another ten years. In 1963, the California State Bureau of Food and Drug Administration indicted her for grand theft, but Drown died before the state was able to bring her to trial. □

Some Liked It Hot

For thirty years in the early 1800s, a medical treatment that was known as Thomsonianism swept the young United States. Based on heat and herbs, the therapy offered every man and woman the chance to be his or her own doctor.

It began when Samuel Thomson, an untutored son of a New Hampshire farmer, watched his mother die of "galloping consumption" under the care of doctors and saw his wife brought to death's door during childbirth. Desperate, he had called in a root doctor—a specialist in natural herbs—to treat her. She swiftly recovered, and Thomson became both an avid botanic doctor on his own and a gadfly for the medical profession.

He combined Indian and folk cures learned during his rude youth with a dogmatic faith that "heat is life, cold is death." The cause of all illness was cold; the universal remedy, then, must be heat. To ensure that the body's furnace, the stomach, burned at peak efficiency, it had to be swept clear of all "obstructions."

In Thomson's treatment, heat was applied externally by thirty-minute steam baths and internally by such hot botanicals as red pepper. Then the interior was cleared with enemas, purgatives, and such powerful natural emetics as lobelia. Though not for the fainthearted, the arduous therapy was often preferable to more conventional techniques, which doctor-hating Thomson called "their instruments of death, Mercury, Opium, Ratsbane, Nitre, and the Lance."

After honing his methods on his children and his neighbors, Thomson toured New England, dosing patients with steam, lobelia, and sixty-four other roots and herbs, and decrying such questionable practices of the established medical community as bleeding and mineral tonics. In 1809, Thomson was arrested for the murder of a patient, but he was acquitted when his lawyer ate the supposedly poisonous plant without any ill effects. (A botanist subsequently testified that the plant introduced into evidence as lobelia—the putative agent of death—was harmless marsh rosemary.)

Although two patents were granted to Thomson, his treatments gradually fell out of fashion. While doctors continued to explore the medicinal value of herbs, Thomson's simple heat-is-life theory, and the so-called steam doctors who practiced it, disappeared. But Thomson followed his dogma to the end and was treating his own afflictions with herbs and steam when he died in 1843. □

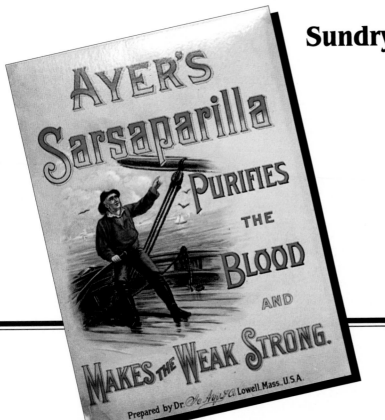

Sundry Tonics and Elixirs

Quack remedies of the late 1800s promised relief for every ailment from alcoholism to "paralysis, softening of the brain, and mental imbecility," as the huckster of Moxie Nerve Food declared. The key to Moxie, the formulation of Dr. Augustin Thompson of Lowell, Massachusetts, was supposedly a South American plant resembling asparagus or sugar cane; its identity remained a proprietary secret.

The popular nerve tonic Paine's Celery Compound contained leaves from a less mysterious Andean plant, coca. Its other main ingredient: 21 percent alcohol. ◊

Millions of people swore by the healthful effects of sarsaparilla, a root-beer-like soft drink that was prepared from a tropical member of the lily family. One brand of sarsaparilla, called Panacea, was billed as a potent tonic and garnered loyal customers nationwide. Taking advantage of the shortage of doctors in the expanding West, a promoter named James Cook Ayer amassed a fortune in sales of his sarsaparilla extract.

The Wild West figured in the popularity of another concoction, Kickapoo Joy Juice—later immortalized in the "Li'l Abner" comic strip as the most potent moonshine distilled in Dogpatch. The original was marketed in New England at shows that featured Sioux, Pawnee, Mohawk, Iroquois, and Blackhawk Indians.

For some entrepreneurs, however, such nostrums were mere steppingstones to greater things. In order to promote his Oxien Tablets, made from "the fruit of the Baobab tree," Maine merchant William Gannett issued his own magazine, called *Comfort*. In 1921, his son plowed the publication's profits into a new venture that is today the giant Gannett Company, owner of more than a hundred newspapers—among them *USA Today*—as well as major television and radio stations. □

Pink Lightning

In the late 1880s, Prussian immigrant William Radam believed he had found a kind of universal cure—and a fortune—in a "microbe-killing" tonic that was 99.381 percent water. His search for such a panacea had begun years earlier, when he was debilitated by malaria, rheumatism, and grief over the deaths of two of his children.

At the time, Radam owned a thriving plant nursery and seed store on a thirty-acre farm located near Austin, Texas, where he applied his horticultural skills to finding an all-purpose medicine. He had heard that microbes could cause disease and reasoned that killing them would do for the body what killing insects did for plants. But his early tests only confirmed that most poisons capable of killing fungi tended to kill the host.

Believing that lightning played a sterilizing role in the atmosphere, Radam built a contraption to simulate what he believed were its chemical effects, filling his invention with water and a mixture of sulfur, sandalwood, manganese oxide, nitrate of soda, and other substances. Actually a stove inside a sealed tank, the device produced a brew Radam mixed with wine and sold as a pink elixir called Microbe Killer. Said to contain hydrochloric and sulfuric acid in small quantities, the watery drink became so popular that others counterfeited it. Radam, who claimed himself cured by his pink lightning, was able to leave his Texas farm for a millionaire's mansion on New York's Fifth Avenue. □

Radioactivity: A Lethal Medicine

Although today there is a general fear of radiation from nuclear power plant accidents and such natural sources as the seepage of radon, a gaseous product of radium, radiation has not always been viewed with apprehension.

When the French physicists Pierre and Marie Curie announced their discovery of radium in 1898, there were many who welcomed the strange new element. Its peculiar rays—the stream of particles released in the process of radioactive decay—were widely perceived as a beneficial emanation, "accepted," one doctor averred, "as harmoniously by the human system as is sunlight by the plant."

Entrepreneurs quickly put radium to work. Homeopaths offered powders and tablets dashed with the substance. Druggists peddled radium toothpastes, face creams, and hair tonics. The Q Ray Electro Radio-Active Dry Compress, a heating pad, had a nuclear heat source of decaying radium. The Lifestone cigarette holder filtered the smoke through radium-laced powder to "protect users from lung cancer."

The Revigator, a ceramic crock filled with radium-bearing ore, produced radon-instilled water that was reputed to rival sought-after mineral and spring waters. Instructions on the crock recommended drinking six or more glasses daily. Radithor, a bottled radium water, had its fans too, including an amateur golf champion by the name of Eben Byers, whose two-bottle-a-day habit resulted in a rotting jaw, brain abscesses, anemia, and ultimately, death.

Similarly, laboratory researchers exposed to radioactive substances began to show anemia, bone decay, and cancer—all symptoms of radiation poisoning. What may have been the definitive proof of radium's dangers came in the 1920s. An abnormally high mortality among female painters of luminous watch dials had been attributed to poor hygiene by their employer, the Radium Corporation of America. But further study revealed that the women were the victims of their craft. Before dipping their tiny brushes into radium-laden paint, they pointed the brush tip with their lips—and swallowed deadly traces remaining from the previous stroke.

As evidence of radium's toll mounted, such products and practices largely disappeared. Nevertheless, as late as 1965, one Florida manufacturer was still producing radon crocks. □

The Durable Mrs. Pinkham

For more than ninety years, the kindly, grandmotherly visage of Lydia E. Pinkham peered at customers from bottles of Pinkham's Vegetable Compound. This unlikely figure became an almost universally known celebrity salesperson, the titular head of a multimillion-dollar patent medicine empire that persisted even after Mrs. Pinkham's death in 1883.

One of twelve children from a well-to-do Lynn, Massachusetts, Quaker family, Lydia Pinkham ardently supported the cause of women's rights. This interest took what was to prove a profitable turn in the 1860s, when she began bottling a home treatment for female complaints such as menstrual cramps and fallen uterus. Her basic recipe derived from an old herbal concoction, the secret of which had been given to her husband, Isaac, in partial payment for a debt. Mrs. Pinkham's version contained substances such as unicorn root, black cohosh, and fenugreek seed, but its prime constituent was alcohol—18 percent.

Pinkham's empire was launched in 1875 when her sons Dan and Will encouraged her to sell her medicine. Despite its pungent, lingering taste, Pinkham's compound proved to be a great success. In one of the first instances of saturation advertising, the Pinkhams blitzed newspapers with full-page notices that were designed with banner headlines in imitation of news stories.

Employing what would become known as the "pain and agony" pitch, these notices chronicled the physical travails of women and offered testimonials for the compound as well as increasingly exaggerated health claims in small print. The advertisements, which capitalized on women's fears of unsympathetic male doctors, were supplemented by pamphlets, drugstore displays, billboards, and even greeting cards, most of them bearing the benign image of Lydia Pinkham.

For a number of years, Mrs. Pinkham answered a booming correspondence from women who claimed to have been helped by her compound; their cures became fodder for more advertisements. The company continued to exhort readers to "write Mrs. Pinkham" even after her death in 1883, having hired a full-time staff of women to respond to the letters. This questionable practice ended, however, following an exposé by the *Ladies' Home Journal* in 1905. Sales of Pinkham's compound continued to be brisk, peaking in 1925. But the public's changing tastes, Food and Drug Administration bans on medical claims, and endless infighting among the founder's heirs led to an inevitable decline. Pinkham's was sold in 1968 to Cooper Laboratories and closed down in 1973, nearly a full century after the first bottle of tonic had been purchased. The Pinkham legend has lived on, however. To this day, there are a number of residents of Lynn who attribute the longevity of their neighbors' lives to Lydia E. Pinkham's Vegetable Compound. □

Happy Days

Hadacol, an elixir that became almost an institution in postwar America, was the brainchild of Dudley J. LeBlanc, the son of a Cajun blacksmith, born in the bayou country of Louisiana in 1895. A shrewd salesman from an early age, LeBlanc started out selling shoes and tobacco while dabbling in politics and the burial insurance business on the side.

In 1932, after serving in the state senate for six years, LeBlanc made a bid for the governor's office. When he lost to former governor Huey Long's candidate, he abandoned the burial insurance business and began manufacturing patent medicines called Dixie Dew Cough Syrup and Happy Day Headache Powder. In 1941, his supply was seized and destroyed by Food and Drug Administration agents who charged that LeBlanc had promoted his products with false health claims.

But LeBlanc was already on his way to better things. In 1943, an infected toe had put him near death. He later claimed to have been saved by injections of a certain elixir, which he resolved to duplicate. When his doctor refused to reveal the ingredients of the lifesaving potion, LeBlanc pilfered a bottle and analyzed the contents, finding them to be mostly vitamin B-12, with traces of iron, calcium, phosphorus, hydrochloric acid, and honey—spiked to a total of 12 percent with alcohol.

LeBlanc brewed a batch of the stuff behind his barn, bottled it, and sold it as a nostrum called Hadacol, a contraction of Happy Day Company plus *l*, for LeBlanc. Enthusiastic radio commercials amply fortified with testimonials from satisfied users went out in French to what LeBlanc called "the plain-living, hard-to-convince families of [the] romantic delta land."

These were the lucky people, he said, who became the "first to see with their own eyes how this unknown new health formula marches into the battle against the pain

Dudley J. LeBlanc *(below, left)* poured millions into extravaganzas featuring pretty girls, celebrities, and clowns *(below)* to sell Hadacol, a popular potion that made and lost a fortune for him.

and suffering of disease." Hadacol, the ads said, would alleviate rheumatism, heart trouble, and asthma, at $3.50 a bottle.

Southerners flocked to buy the concoction, enabling LeBlanc to extend his sales pitch to other areas. Although enjoined by the Federal Trade Commission from making health claims for his product, LeBlanc was by 1950 engaged in a massive million-dollar-a-month promotional campaign, complete with a touring, old-time medicine show—and he was grossing some

$20 million in twenty-two states.

In order to open the lucrative western market, LeBlanc staged a month-long carnival in Los Angeles featuring such Hollywood luminaries as Judy Garland and Mickey Rooney. At the same time, though, the entrepreneur was careful to mute his health claims for Hadacol. When the comedian Groucho Marx asked him what the stuff was good for, LeBlanc quipped that it "was good for five and a half million dollars a year." The reply was in the spirit of the times, for

Hadacol jokes swept the nation, many of them hinting at the potion's power as an aphrodisiac.

For Louisianans, though, the best Hadacol joke came in 1951, when LeBlanc suddenly sold out to Yankee businessmen. Their $8 million purchase price, of which LeBlanc received $250,000 in cash, bought a set of cooked books, more than $2 million in unpaid bills, and a huge surplus of Hadacol in an already saturated market. The wily LeBlanc had baled out in the nick of time. □

Cruel Cure

Harry Hoxsey inherited from his father—later, he said it came to him from his great-grandfather—an herbal recipe that he parlayed into big business. From Dallas, Texas, beginning in 1936, he built up a chain of facilities that specialized in cancer treatment. For $400, patients received Hoxsey's formula, a potpourri of herbs including red clover blossoms, licorice root, and the bark of the buckthorn, prickly ash, and barberry—and arsenic.

American Medical Association and Food and Drug Administration officials repeatedly charged Hoxsey with quackery, but not until 1960 did the courts enjoin him from practicing. Meanwhile, thousands of desperate people flocked to him for his supposed nonsurgical cure.

National Cancer Institute studies subsequently revealed that Hoxsey's formula actually contained antitumor compounds that may have made it mildly effective against some types of skin cancers. However, arsenic-rich salves, which destroy both healthy and cancerous tissue, were nothing new: They had been used against external cancers since the sixteenth century. And his medicines had no power over internal tumors.

Despite its impotency, Hoxsey's cruel cure lives on in Mexican border town clinics. But Hoxsey himself died in 1974—as his parents had, of cancer. □

Puzzling Remissions

On May 19, 1947, a fifty-nine-year-old man arrived at the George Washington University Hospital complaining of a persistent cough, weight loss, and general fatigue. X-rays and a biopsy showed that the three-pack-a-day smoker had inoperable lung cancer. He was quickly hospitalized. Soon, to his doctors' surprise, the cancer began to reverse itself. The patient was released. When examined five years later, almost no trace of his cancer could be found.

Although detailed studies of such inexplicable spontaneous remissions from cancer rarely appear in the medical literature, this and similar tales testify to the complexity and potency of the human immune system, and to the unpredictable behavior of the disease itself. Physicians are hard pressed to explain such extraordinary recoveries as that of a New York woman whose body was riddled with cancer, but who went into total remission following the sudden death of her detested husband.

Fevers and stress may also play key roles. In the early 1900s, Dr. William Coley of the New York Cancer Hospital achieved thousands of cures by infecting cancer patients with a bacterium that induced high body temperatures. Scientists at the Sloan-Kettering Institute later found the infection stimulated the release of a natural cancer fighter called tumor necrosis factor. And an Australian patient reportedly rid himself of advanced bone cancer by lowering his anxiety level through acupuncture, massage, yoga, and daily meditation.

Unfortunately, the possibility of such remission has helped fuel the sales of ineffective nostrums offered by a legion of cancer quacks. But, in fact, such cures are the exception, not the rule, possibly occurring only once in a hundred thousand cases. Gambling on the body's ability to evict such killers without the aid of medical science remains a very long shot indeed. □

Graveyard Medicine

Since the days of classical antiquity, superstition has held that the bones and body parts of the dead have magical healing powers. In eighteenth-century England, a brisk traffic sprang up that dealt in the fingers, skulls, and hands of hanged men, which were thought to rid the skin of warts and wens. For a nominal fee, the hangman sometimes offered his similarly potent touch. The *Casket*, a journal published in Philadelphia in 1833, reported that snorting powdered moss scraped from decaying skulls would dispel headaches. And as recently as 1960, historians of folk medicine report, the belief that a dead man's touch shrinks tumors and goiters still existed in parts of Pennsylvania. □

I Shall Please

From the Middle Ages to the 1800s, theriac—also known as the calendar drug, since it had more ingredients than a month had days—was a widely used European medicine. In medieval times, the concoction reputedly included ground Egyptian mummy parts and viper flesh. By the time of Claude Bernard, famed French physiologist of the early nineteenth century, the elixir had come to consist of more common materials.

During a youthful stint working in a druggist's, Bernard observed the making of theriac: Into a large jug of water that sat ever ready on the counter went any old or surplus prescription. Ladled out to buyers, this brew caused no great harm, since any dangerous agents were diluted. But it was generally reputed to do much good.

What powers theriac and similar cure-alls did have were probably due in large measure to the placebo effect. Placebo, from the Latin "I shall please," came into English through the Catholic church and in secular usage came to signify flattery. In medical circles, it refers to a substance that has no real chemical action but serves to please patients. Responding to the widely documented placebo effect, patients given a dummy medication—often nothing more than a sugar pill—believe they are being helped, and either experience a diminution of pain or symptoms, or recover completely.

A 1955 study of more than a thousand patients found that of those given placebos, more than a third reported relief, especially from pain and nervous disorders. Placebos have also been effective against heart problems, arthritis, migraine headaches, acne, postoperative pain, warts, colds, cancer, multiple sclerosis, and more.

To be sure, the placebo effect can occur only when a patient has a positive attitude, believing that the supposed cure will work. Conversely, attitude can also play a negative role, producing an effect called nocebo. What may be the classic example of nocebo was that of a patient identified only as Wright. In 1957, suffering from lymphatic cancer and tumors the size of oranges on his body, Wright begged his doctor for a new drug called Krebiozen, which was supposed to be effective against his kind of cancer. A week and a half after his doctor injected him with the experimental drug, Wright's tumors disappeared.

But when Wright read a newspaper report that Krebiozen had failed clinical tests, his tumors returned. Noting the psychological effect, Wright's doctor then administered what was supposed to be a new, potent form of the drug but was in fact water. Again, Wright's tumors vanished. Two months later, the definitive American Medical Association report on Krebiozen declared it worthless for lymph cancer. Wright died two days after reading of this study.

Certain that such happenings are due to more than just the power of suggestion, some doctors have urged that their colleagues reexamine the nature of the doctor-patient relationship, which may work in a manner similar to the placebo effect. For example, some patients appear to heal better if their doctors have shown enthusiasm for a certain drug or course of therapy, or have discussed strategies for treatment in detail. □

Visitors Only

In 1858, the French peasant girl Bernadette Soubirous—later to become a saint—claimed that she had seen an apparition of the Virgin Mary near the town of Lourdes. Ever since, throngs of believers have made a yearly pilgrimage to the Pyrenees village in search of miraculous cures.

A distinguished board of scientists appointed by the Roman Catholic church scrutinizes all claims of healings from Lourdes, separating the persuasively miraculous from the easily explained. Of the hundreds of thousands of "cures" that have taken place since Bernadette Soubirous's vision, the Church recognizes only sixty of them as true miracles.

But the scientific board has found something else as well: The power of Lourdes seems to work only on tourists. Except for a few early cases, not a single cure has been claimed by a resident. □

The Geography of Healing

Despite its emphasis on rigorous scientific methods, modern medicine is subject to trends and quirks that vary from nation to nation. Different government approaches to the funding of health care, bureaucratic roadblocks, the influence of drug companies, rivalry among local doctors, the fears and biases of patients, and other factors that are not directly related to the healing art combine to determine how any given ailment will ultimately be treated.

Whereas the Japanese have an aversion to being operated on, they consume more prescription drugs per capita than any other people. This is partly due to a system in which doctors, not pharmacists, dispense drugs—at a considerable markup. The more drugs the doctors decide should be doled out, the greater their profits. Americans also use large quantities of drugs. Their hospitals, for example, administer about twice as much medicine as a Scottish one would for the same illness.

Although Britain has a higher incidence of coronary heart disease than does the United States, doctors there perform only one-sixtieth as many bypass operations per 100,000 people. American physicians opt for Cesarean delivery twice as often as their British counterparts do, perhaps fearing malpractice suits if a natural delivery goes awry.

Some cultures show marked preferences for certain drugs. Spaniards take four times as many antibiotics as the French, who take three times as many tranquilizers and sleeping pills as Spaniards. Physicians in seven European countries that have a similar incidence of diabetes show markedly different tendencies in the prescribing of antidiabetic drugs.

Japan spends more on health care per individual citizen than any other nation. Medical costs in the United States are among the highest in the world, yet child mortality, a traditional gauge of public health, outstrips that of all other industrialized nations. Similarly, Ireland, which spends a greater share of its gross national product on health care than any other European Community country, has the shortest life expectancy for males. But Greece, which spends the least, has the longest. □

THE BODY TEMPORAL

The complex dance of human life follows a number of cycles, so varied in nature and duration that their sum seems merely chaotic. Disparate as they are, these rhythms nonetheless order the body's existence, from the first fluttering beat of a baby's heart to the inevitable physical decline that comes at the close of a long life.

Some of the rhythms synchronize people: For example, waking and sleeping are obviously linked to night and day. But there are also other, less apparent patterns that seem to have their own timetables, some affecting the entire species, others highly individualized.

In addition, statisticians analyzing census data encounter unexpected tempos that seem to vary from race to race and from one location to another. Biologists have discovered hormonal tides that regularly ebb and flow through the body. And medical researchers have found that disturbances to these bodily rhythms can shake the entire edifice of physical and emotional health.

To detect such designs, however, is not necessarily to understand them. Although the temporal qualities of life have become the focus of a new scientific discipline called chronobiology, researchers are still a long way from explaining the complicated interplay of time and the body.

A Clockwork Organ

Waking in predawn darkness, a man turns toward his alarm clock moments before it sounds. Reaching to turn it off, he may briefly question the need for a mechanical device when his internal timekeeper works so well on its own.

This circadian—or twenty-four-hour—chronometer is familiar to everyone who has had to meet a schedule when no alarm clocks were at hand and was long thought to be purely mental. But recent studies indicate that the biological clock is in fact physiological. Apparently, the body's built-in chronometry comes from the so-called SCN, or suprachiasmatic nucleus, a cluster of neurons sitting astride a fluid-filled space in the midline of the brain.

In studies with rats conducted in 1972, researchers learned that damage to the SCN could disrupt daily rhythms of water drinking and physical activity. By the early 1980s, similar effects had been observed in cats, squirrels, and monkeys. Further experiments showed that transplanting an SCN from one guinea pig into the brain of another gives the recipient the donor's time schedule.

Despite the lack of such experimental evidence for humans, researchers assume the human SCN, which resembles those of other mammals, performs similarly. Its functions might include regulating daily cycles such as waking and sleeping, timing heart and pulse rates, and governing the secretion of hormones and enzymes—including those from the pineal gland *(page 121)*—that in turn control many organs throughout the body.

Although the SCN is probably not the body's sole internal clock, it may be both the most important and the most compact: This exquisitely complex biochemical metronome is concentrated in a bunch of brain cells that are smaller than a grain of sand. □

Primordial Snooze

Continuing a practice he had observed since his days as a young army officer in India more than forty years earlier, Winston Churchill almost always got an hour of sleep sometime between three and six in the afternoon, even when German bombs rained on London at the height of the World War II Blitz.

The prime minister's urge to snooze was attributed by many to the substantial midday meal and generous ration of wine and spirits that were also part of his routine. But more recent study indicates that Churchill was instead yielding to an afternoon drowsiness that appears to be part of the natural sleep cycle, regardless of the time or size of lunch, amount of drink, or geographical location.

This universal sleepiness is reflected in human performance. Factory workers make more mistakes in the midafternoon, office workers begin to daydream, and schoolchildren stumble through arithmetic problems. In many parts of the world, custom recognizes this condition and allows for an early afternoon nap, or siesta.

According to New Orleans anthropologists Marcia Thompson and David Harsha, the urge to nap has its roots in prehistoric experience. Thompson and Harsha point to the behavior of monkeys and apes, who generally spend their mornings and their late afternoons in a noisy bustle of communal feeding but lapse into quieter activities during the hottest part of the day. For a few relaxed hours, they play, groom each other,

sprawl idly, and take short naps.
Early bands of hominids probably engaged in similar activities before migrating from the tropics, and the age-old habit has persisted through human evolution. Thus, under the war-torn skies of England, Churchill may have napped to a beat laid down by the African sun a million years ago. □

Graveyard Shift

"In the real dark night of the soul," wrote the American novelist F. Scott Fitzgerald, "it is always three o'clock in the morning." For humans, the early morning hours are not merely depressing—they also mark a period of minimum performance. Body temperature and blood pressure drop to their lowest daily levels, and the brain's ability to handle such complex tasks as mathematics dwindles.

Most people sleep through this predawn low. Those who do not— even workers who customarily sleep by day and work by night—often pay a price. More accidents happen at night, and they tend to be more serious than their daytime counterparts. One study shows that almost 40 percent of night-shift factory workers report errors, near misses, and accidents that are relatively rare on the day shift.

And some investigators believe that the 1979 nuclear reactor malfunction at Pennsylvania's Three Mile Island might not have been so bad had it not occurred at four in the morning. When the alarm sounded, power-plant workers had to make critical decisions while their own faculties were at their lowest ebb. Human errors compounded mechanical failures, permitting what should have been a minor incident to develop into a serious mishap. □

Cave Dweller

On January 13, 1989, Stefania Follini settled down to live in a cave for a few months. An interior designer by trade, the diminutive twenty-seven-year-old Italian had volunteered to serve as an experimental subject in a study of the effects of timelessness. Her new abode, a sparsely furnished, room-size box with clear plastic walls and ceiling, was in the Great Hall of Lost Cave, thirty feet below the desert near Carlsbad, New Mexico.

One point of the $200,000 experiment was to measure the psychological and physiological effects of cutting off all clues to the passage of time, such as clocks and the rising and setting sun. A support team on the surface monitored her timeless life in a nether world constantly illuminated by electric lighting. Although microphones and video cameras captured her every sound and move, her only two-way communication with the surface was through a personal computer.

Researchers already had an idea what to expect. In similar experiments, subjects had tended to settle into a pattern of waking and sleeping that was longer than the ordinary, sun-regulated day. The average length of those cycles, from waking to waking, was about twenty-five hours, although some people developed rhythms around "days" that stretched as long as forty-eight hours.

Follini proved no exception. A few days after entering Lost Cave, she began to shift to a daily cycle about twenty-eight hours long, but maintained a regular routine. Upon rising, she took her blood pressure and a urine sample, then exercised before breakfast. Reading, drawing, and playing her guitar consumed the hours until lunch, after which she read and took a nap. Follini would get more exercise practicing judo (she holds a brown belt) or exploring the cave before supper. Then she read and went to bed. At intervals she performed additional medical tests

and communicated with the support crew above.

Six weeks into the experiment, Follini's rhythms underwent a dramatic change. Awakening at one in the morning, she did not go back to sleep for twenty-four hours. Five days later, she had a waking period of almost thirty hours. She had shifted to a forty-four-hour cycle and lost most perception of the passing of time.

The change was imperceptible to Follini, whose schedule of activities remained the same. In the course of a seven-hour interview (conducted through her computer) she noted, "I am sleeping a lot, but this is normal for me. I sleep well. I sleep frequently." Follini's misjudgment of elapsed time was not limited to sleep: At the end of the interview, she estimated that it had taken about an hour.

Weathering bouts of mild depression—she sometimes found herself conversing with visiting mice and frogs, the original tenants of the cave—Follini lasted until the research team called an end to the experiment on May 22, after 130 days. The news confused her when it flashed across the computer screen. "How come?" she typed back. Follini thought she still had many weeks to go, for her tally of the long subterranean days put the date at only March 14. Time, she said when she returned to the surface, had become "a continuous moment." □

The Blahs of Winter

In the middle and northern latitudes of the United States, about two people in ten experience a kind of winter doldrums. Their spirits fall with the shortened days of winter. They tend to put on pounds easily and have a hard time getting up in the morning. Nonetheless, most make it through the day, and through the winter, with minimal difficulty. But, for about a quarter of this group, the blahs take a more serious turn into what is called seasonal affective disorder, or, aptly, SAD.

As winter deepens, the appetites of those affected by SAD may become voracious, causing large ◊

Above the Arctic Circle, Tromsø, Norway, sees no sunlight for two months during the winter. Following the lead of experiments linking darkness and depression *(inset)*, residents combat the blahs with heavy doses of artificial light.

weight gains. Depressed and increasingly withdrawn from the world, some sufferers become bad-tempered and may sleep an extra four hours or more each day. Only the lengthening days of spring free them from the growing burden of their malady.

The disorder may well be an extreme form of a seasonal cycle inherited from the earliest migrants to northern climes. According to Frederick Sargent of the University of Texas, those human ancestors may have reduced their amount of activity in winter to a kind of semihibernation, sleeping more and burning up the body fat accumulated from eating lots of carbohydrates during the summer and the autumn.

The best treatment for SAD symptoms may be the simplest one. Research at the National Institute of Mental Health in Washington, D.C., shows that the spirits of the afflicted can be lifted by a simulated spring of bright, full-spectrum light, applied for a few hours each day. □

Jetless Lag

The first traveler heard to complain about the debilitating effects of crossing time zones—what is today known as jet lag—was not a jet-setter but American aviation pioneer Wiley Post. Before his path-breaking flight around the world in 1931, the eye-patched Oklahoman spent a number of months sleeping at different hours each day in order to disrupt his normal patterns of rest.

"I knew that the variance in time as we progressed would bring on acute fatigue if I were used to regular hours," he wrote after the successful flight, which took eight days, fifteen hours, and fifty-one minutes from start to finish at Roosevelt Field near New York City.

As it happened, Post's preparations were of little avail. He and his navigator, Harold Gatty, were already suffering from exhaustion by the time they reached Germany on the second day of the trip. In fact, Post was so tired, and consequently forgetful, that he was forced to return to Hanover immediately after starting a one-hour leg to Berlin: He had taken off with nearly empty fuel tanks in his propeller-driven, 170-mile-per-hour monoplane, *Winnie Mae.* □

Light for the Lagged

Ever since jet lag began to plague large numbers of long-distance fliers during the mid-1960s, self-styled experts have advanced such "cures" as strict dietary regimes, herbal concoctions, and mild sleeping medications. But none of these remedies has produced convincing results.

In the late 1980s, however, a group of American scientists proposed a novel—and seemingly effective—technique for rapidly resetting human clocks. Their method is based on the fact that the body takes many of its time cues from light. By exposing experimental subjects to an array of bright lamps for a period of several hours each day, they effectively recalibrated the volunteers' internal timepieces. As one of the research leaders, Harvard professor Richard Kronauer, told a reporter in 1989, "I could set you to Bombay time after just two or three treatments"—a ten-hour fast forward from local time at Kronauer's Boston-area sleep clinic.

To be effective, the lights must be as bright as daylight and ad-ministered at just the right point in the cycle of waking and sleeping. These brilliant pulses trigger reactions within the brain that have the effect of stopping the internal clock, then starting it up again at the desired time. The results, according to Kronauer, can be as precise as the adjustment of an alarm clock.

Because few people have access to special lights such as those used in the Harvard study, Kronauer devised a set of rules for using ordinary daylight and darkness to reset the internal clock by as much as three hours each day. The rules, based on a mathematical formula derived from study data, are almost identical for eastbound and westbound travel: Avoid daylight in the morning but get as much as possible in the afternoon and early evening.

A traveler on an evening flight from New York to London, for example, should keep away from the morning light on departure day and after arrival, then spend most of the first afternoon in London outdoors. A passenger on a day flight in the other direction should also avoid the morning sun but sit by a window to get maximum benefit from the afternoon light.

Kronauer and his associates use the bright-light technique whenever they travel, but most people remain unaware of this innovative treatment. There may, however, be a few peculiar side effects. For example, one faithful practitioner found himself explaining his jet-lag cure to suspicious Heathrow airport security personnel, who took him aside to inquire why he had arrived in London wearing a pair of welder's goggles. □

Love in a Cold Climate

Among the Inuit Eskimo of frigid Labrador, the birthrate is firmly tied to the calendar, but not as one might expect. Instead of high births occurring nine months after the cold, dark interval of the Arctic winter, Inuit births peak in March, then fall to a minimum during summer. The seasonal variations of birthrates there are four times those in other parts of the world.

Some observers believe that cultural taboos are responsible for the improbable cycle; others suggest that extreme winter cold can be blamed for the dearth of summer babies. But Joel Ehrenkranz, an American physician who has studied the phenomenon, discounts these explanations. The pattern has persisted through two centuries of extreme cultural upheaval, he notes, adding that "even in the dead of winter, Inuit houses aren't all that cold."

Ehrenkranz found a hint of the answer in a report by F. A. Cook, a New York surgeon who wintered in Thule, Greenland, in the 1890s. During the four-month Arctic

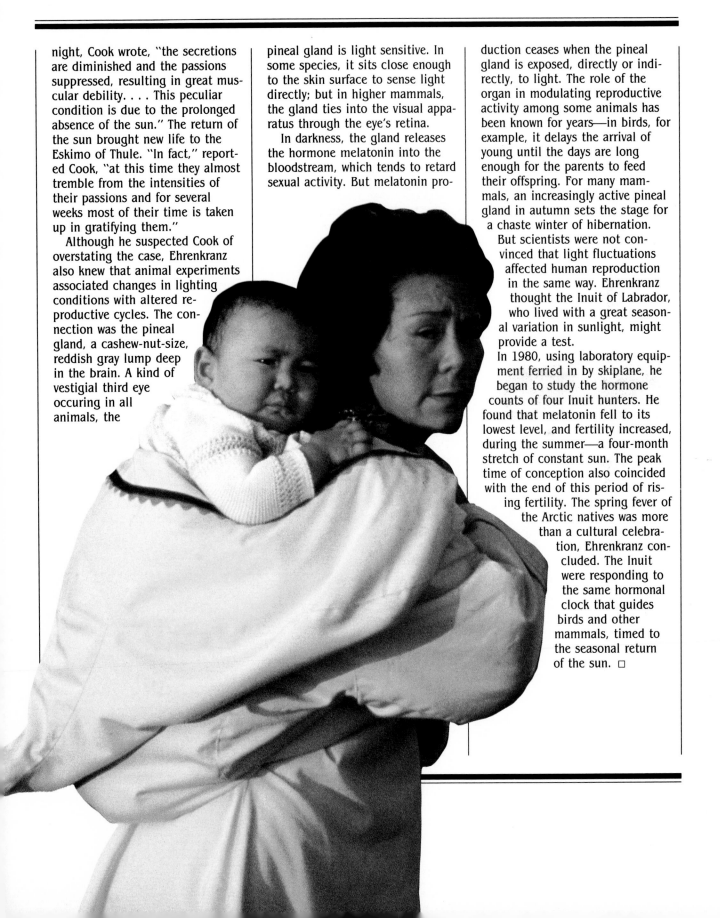

night, Cook wrote, "the secretions are diminished and the passions suppressed, resulting in great muscular debility. . . . This peculiar condition is due to the prolonged absence of the sun." The return of the sun brought new life to the Eskimo of Thule. "In fact," reported Cook, "at this time they almost tremble from the intensities of their passions and for several weeks most of their time is taken up in gratifying them."

Although he suspected Cook of overstating the case, Ehrenkranz also knew that animal experiments associated changes in lighting conditions with altered reproductive cycles. The connection was the pineal gland, a cashew-nut-size, reddish gray lump deep in the brain. A kind of vestigial third eye occuring in all animals, the

pineal gland is light sensitive. In some species, it sits close enough to the skin surface to sense light directly; but in higher mammals, the gland ties into the visual apparatus through the eye's retina.

In darkness, the gland releases the hormone melatonin into the bloodstream, which tends to retard sexual activity. But melatonin pro-

duction ceases when the pineal gland is exposed, directly or indirectly, to light. The role of the organ in modulating reproductive activity among some animals has been known for years—in birds, for example, it delays the arrival of young until the days are long enough for the parents to feed their offspring. For many mammals, an increasingly active pineal gland in autumn sets the stage for a chaste winter of hibernation.

But scientists were not convinced that light fluctuations affected human reproduction in the same way. Ehrenkranz thought the Inuit of Labrador, who lived with a great seasonal variation in sunlight, might provide a test.

In 1980, using laboratory equipment ferried in by skiplane, he began to study the hormone counts of four Inuit hunters. He found that melatonin fell to its lowest level, and fertility increased, during the summer—a four-month stretch of constant sun. The peak time of conception also coincided with the end of this period of rising fertility. The spring fever of the Arctic natives was more than a cultural celebration, Ehrenkranz concluded. The Inuit were responding to the same hormonal clock that guides birds and other mammals, timed to the seasonal return of the sun. □

Cyclemania

Creative activity in human beings appears to rise and fall in a cycle that lasts 7.6 months from peak to peak. Physical strength may fluctuate with a twenty-three-day rhythm, somewhat faster than the thirty-five-day tempo of emotions found among working men. Mental activity (measured in terms of patent applications and library withdrawals) and violent crime both seem to ebb and flow over a period of 12 months.

Humans have long understood such basic rhythms of life as waking and sleeping, planting and harvesting, but it remained for statisticians to uncover the more obscure cadences. One man, Edward Dewey, tracked down scores of such natural patterns, some real, some spurious, and made a life's work of trying to understand the cyclical phenomena that seemed to pervade the universe.

Dewey, a Harvard-educated economist, discovered his vocation while serving as the chief economic analyst for the U.S. Department of Commerce. He had taken his first government job just a few weeks before the 1929 stock market crash, and his search for an explanation of the subsequent economic depression led him to a theory of business cycles.

Upon further study, Dewey found that repetitive events occur in nature as well as in business, and he wondered whether they were connected. To investigate, he established a foundation, attracting a number of eminent scientists to its executive committee. Dewey and his colleagues discovered—among several other things—that the population of Canadian lynx peaks every 9.6 years; that voluntary payments of delinquent taxes in Britain reach a high every 3.5 years; and that a 4-year cycle applies to cheese consumption, plankton yields, pork prices, and the population of ruffed grouse.

Dewey maintained a broad curiosity that eventually swelled to a fixation. By the 1970s, he had embarked upon a quest for a single explanation of all cycles. At first he thought the answer lay in the sunspot cycle of 22.7 years, a thesis he later discarded in favor of one involving the changing alignment of planets as they move in their orbits around the sun.

But the man who founded the study of cycles eventually realized that he could not achieve the deeper understanding he sought. More information was necessary to explain irregularities and to disclose subtle links between patterns. Looking to the future, he invoked the great astronomers who first explained the dynamics of planetary movement: "What we need is a Copernicus and a Kepler in cycle study."

The foundation carried on Dewey's work after his death in 1978, but no grand theory has emerged that can reconcile the thousands of documented cycles. Indeed, it may be that despite the fact that they wax and wane every 18.2 years, there is actually no link between the flood stages of the Nile River and the number of marriages in St. Louis, Missouri. □

Just Five Score and Ten

For individual human beings, the longest cycle is that of a single lifetime. Thousands of years ago, the Book of Psalms reckoned the standard interval between birth and death at three score years and ten—seventy years.

The figure has remained remarkably constant. Even in modern developed nations, the average lifetime lasts only seventy-six years—although a far higher proportion of today's population succeeds in attaining the biblical ideal.

Most people born in ancient societies, afflicted by plagues, pestilence, famine, and war, were lucky to live into their thirties. But those who survived childhood diseases and the rigors of young adulthood and maternity had at least as good a chance as any modern to reach a ripe old age. The psalmist allowed that "by reason of strength" some might live to eighty, and many did, including such Greek luminaries as Plato, who reached the age of eighty-one, Sophocles, who lived to eighty-nine, and Hippocrates—a medical

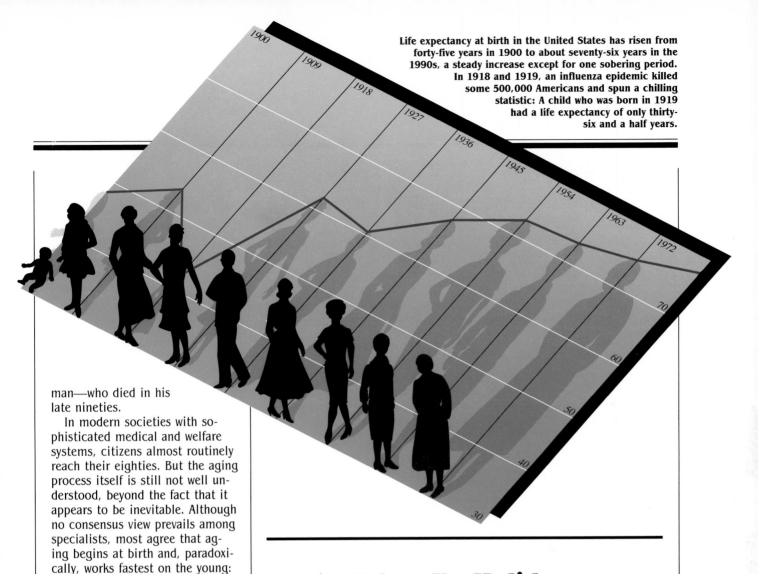

Life expectancy at birth in the United States has risen from forty-five years in 1900 to about seventy-six years in the 1990s, a steady increase except for one sobering period. In 1918 and 1919, an influenza epidemic killed some 500,000 Americans and spun a chilling statistic: A child who was born in 1919 had a life expectancy of only thirty-six and a half years.

man—who died in his late nineties.

In modern societies with sophisticated medical and welfare systems, citizens almost routinely reach their eighties. But the aging process itself is still not well understood, beyond the fact that it appears to be inevitable. Although no consensus view prevails among specialists, most agree that aging begins at birth and, paradoxically, works fastest on the young: An infant ages far more in a year than an adult. Thus its pace slows, but the process of aging continues unabated until death.

The burden of years does not alone bring life to a close, however. With the passage of the years, the body slowly loses its ability to rebound from disease and injury. The elderly succumb when their bodies can no longer rise to the challenge of survival.

Advances in medical science have helped shore up the health of older people and, in doing so, prolonged their lives, often by what physicians refer to as "heroic" means. But nothing yet discovered can extend human life much beyond the age of 110, which appears to be a natural limit. □

Death Takes No Holidays

Part of the justification for disease research is the notion that the elimination of any cause of death will automatically lead to an increased human life expectancy and diminished mortality.

Cancer, for example, kills nearly 500,000 people each year in the United States alone. It would seem, therefore, that a cure would make a substantial dent in the overall death rate and consequently increase the life expectancy of the average person. In fact, many of the people rescued from cancer are at risk from other diseases, which soon begin to take their toll. According to the calculations of some demographers, completely vanquishing cancer would finally translate into only a 2.8-year increase in life expectancy.

In reality, it appears that no amount of curing can greatly alter the dimensions of human mortality. Researchers at the University of Colorado note that as such infectious diseases as tuberculosis and cholera were overcome, degenerative illnesses such as cancer and heart disease moved in to replace them. Scientists speculate that triumph over these disorders will be canceled out in turn by new killers, borne not by bacteria or old age, but by diseases linked to changes in environment and human behavior. □

Tom Parr

The most widely famed claimant to an extremely long life was probably Thomas Parr, an English farmer often cited as the oldest man of the modern era. Reportedly, he was born in 1483; although documentation of his early years is sparse, he gained renown as "Old Parr" in his native Shropshire, where he was said to have worked his fields continuously until his eyesight began to fail him at the age of 130. Parr outlived his first wife—whom he married at 80 years of age—and he was pleased to recall that he had sired an illegitimate child in his 106th year.

It was Old Parr's misfortune to be discovered in the spring of 1635 by the earl of Arundel, who was inspired to set this curious geriatric example before King Charles I. Transported to London in a litter and presented at court, Parr was made a royal protégé under Arundel's care and given what became a death sentence of London living.

More than a century spent in the clear air of Shropshire had ill-prepared Parr for the filthy, smoky capital or its excesses—which for him included rich food, drink, and late hours. On November 14, 1635, claiming an age of 152, Parr died of what all agreed was a disastrous change of living habits. As the royal physician, William Harvey, wrote after his postmortem examination: "Had nothing happened to interfere with the old man's habits of life, he might perhaps have escaped paying the debt due to nature for some little time longer." □

Old Frauds

Rare as it is for anyone to reach the age of 100—fewer than .0001 percent of Americans live so long—there are some places in the world that seem to teem with centenarians. These tend to be isolated, impoverished communities in mountainous farming regions.

In 1972, Vilcabamba, a village of 819 souls high in the Ecuadoran Andes, claimed a total of nine centenarians. Only the year before, the death of a villager by the name of José Toledo Avedano had been officially recorded as occurring at the age of 140. In 1977, Gabriel Erazo Aldean passed on, reputedly aged 132.

Unfortunately for Vilcabamba's reputation, the local church had relatively complete records of births in the village. In the late 1970s, two American researchers examined this baptismal log, using the information to assemble complete genealogies for each of the supposed centenarians. Publishing the results of their study in 1979, Richard B. Mazess and Sylvia H. Forman concluded that none of the village elders had actually reached the age of 100.

Mazess and Forman found that the people of Vilcabamba began to exaggerate their age when they entered their sixties. In the typical case of Miguel Carpio Mendieta, who was born in 1884, the researchers found records from 1944 in which he reported his age as 70. Five years later, he claimed to be 80. By 1970, when he was actually in his late 80s, he claimed an age of 121. Near death six years later at 92, he put his age at "about 130"—which gave him a supposed birth date five years before his mother's.

The American researchers noted that age exaggeration seems to be common among the extremely old around the world, particularly in areas where there is high illiteracy and little formal documentation. The motive for such deception, Mazess and Forman concluded, is the high social prestige afforded to the aged, who are respected for their wisdom and their experience and perhaps for the fact that they have outlasted so many of their contemporaries. □

Togetherness

Ernie Scott didn't get nearly as much attention for his hundredth birthday as he and his wife, Maud, ninety-seven, got on their anniversary a year later. Married in South Dakota in 1909 and still together in 1989, they were the world's longest-wed living couple.

Their eighty-year union had survived plenty of ups and downs: frequent moves around the Midwest, the births of three children, and numerous separations when Ernie was on the road working as a ranch hand. In 1965, they finally settled in Gilroy, California, to live out the rest of their lives.

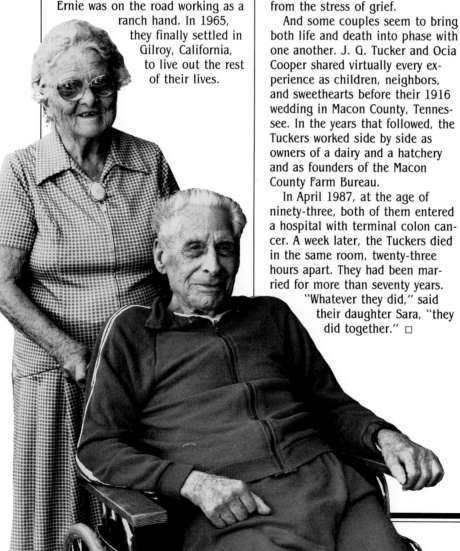

The Scotts are one example among many suggesting that a successful marriage prolongs life, bolstering both parties against the shocks of grief and illness. For example, a 1988 study of changes in life span among the parents of 2,518 Israeli soldiers killed in the 1973 October War found that the loss of a son had no significant effect on parental death rates where the parents were still together. But widowed or divorced parents showed a greater risk of dying prematurely, presumably from the stress of grief.

And some couples seem to bring both life and death into phase with one another. J. G. Tucker and Ocia Cooper shared virtually every experience as children, neighbors, and sweethearts before their 1916 wedding in Macon County, Tennessee. In the years that followed, the Tuckers worked side by side as owners of a dairy and a hatchery and as founders of the Macon County Farm Bureau.

In April 1987, at the age of ninety-three, both of them entered a hospital with terminal colon cancer. A week later, the Tuckers died in the same room, twenty-three hours apart. They had been married for more than seventy years.

"Whatever they did," said their daughter Sara, "they did together." □

A Cruel Acceleration

Every summer, Jason Ellison and about fifteen other children would come together for a reunion, sometimes at a resort, once at Disneyland. Like most childhood treats, the week-long holidays ended all too soon. But for these youngsters, the swift passage of time that paced such pleasures was a tragic fact of their lives.

Children such as Jason and his friends suffer an affliction that terribly accelerates the aging process. Most succumb as teenagers to cancer, hardening of the arteries, and heart failure—the diseases of old age. Their disorder, called Hutchinson-Gilford progeria syndrome—progeria means "before old age" in Latin—occurs only about once in eight million births; in fact, the syndrome is so rare that doctors have had little opportunity to study it. Fewer than twenty living progerians were reported in the world during 1988.

The symptoms are cruel: Progeric children rarely grow larger than a normal five-year-old and are increasingly thin and frail, encountering some of the illnesses usually associated with the elderly as early as age five. Six-year-olds, with bald heads and prominent veins under parchmentlike skin, resemble grownups ten times older.

According to some researchers, progeria may not be true aging, since certain geriatric symptoms—senility, for example—are not seen. Inside their withered bodies, the children are still young, their joys and sorrows, intelligence and wit, intact.

Keenly aware of their differences

from unafflicted contemporaries, many progerians meet others with the disease for the first time at gatherings such as the summer reunion. That event is arranged and funded by the Sunshine Foundation, an organization that provides dream trips to children with terminal illnesses.

The gatherings, which often feature such activities as swimming, camping, and horseback riding, give the children a rare respite from the social tensions that are produced by their unusual appearance. On these outings, according to University of Florida researcher Franklin DeBusk, the children are like any other kids their age— friendly, lively, funny, and mischievous, sometimes happy, sometimes angry, sometimes sad. With completely normal intelligence and emotions, they simply go on living lives that, for reasons science has not yet discovered, speed by much too quickly. □

Sporting Lives

Major-league baseball players seem to live longer—at least the ones who played in the big leagues around the time of World War I. Those players, according to a study published in 1988, lived to an average age of 70.7 years, about 4 years longer than other men of their generation.

The best hitters and pitchers lived longer, a tendency that might be traceable to greater strength and better conditioning. Left-handed players, however, provide a puzzling contradiction. Traditionally the mavericks of baseball, southpaws both great and obscure died nine months earlier than their right-handed counterparts.

Infielders had the best advantage, with an average lifespan of 76 years. Catchers, on the other hand, tended to die younger than the general population.

The higher death rate for catchers, who generally suffer more physical damage than other players during their careers, corresponds to recent surveys of longevity among professional football players. For men who played 5 years at the grueling professional level from 1925 to 1959, the average age at death was just 61 years. □

Feisty Red Sox southpaw pitcher Robert "Lefty" Grove had a distinguished seventeen-year major-league career. Grove died in 1975, aged seventy-five, having outlived most men of his generation by about a decade.

The Cellular Time Bomb

Because so many bodily processes are involved, it is extremely difficult to tell the causes of aging from its effects. Studies of why and how we grow old are further complicated by the length of human life. Researchers working with rats can observe the full birth-to-death cycle in just four years. With humans, of course, equivalent research would span many decades. Accordingly, a number of investigators have focused on the aging process at the short-lived level of the body's cells.

Some of the most telling results have emerged from experiments conducted by University of Florida cell biologist Leonard Hayflick. Taking living cells from animals, Hayflick cultured them in his laboratory, where they continued to reproduce as they would in the body, each cell dividing periodically to make two. But Hayflick found that in every case, the cells were limited in the number of times they could split. After a set number of generations, they ceased dividing; then, no longer able to reproduce themselves, the worn-out organisms died.

Cells from young animals lasted longest, dividing many times before they stopped; cells from aged animals, on the other hand, divided only once or twice. Embryonic cells from human beings reproduced through about fifty generations, corresponding to some 110 years of life.

According to Hayflick, cellular reproduction is a kind of fuse marking life's span. When cells can no longer divide, the fuse has burned its measured length and the body dies. Thus, aging may be less a matter of wear and tear than of keeping an appointment with death scheduled when life, in the form of embryonic cells, began its temporal journey. □

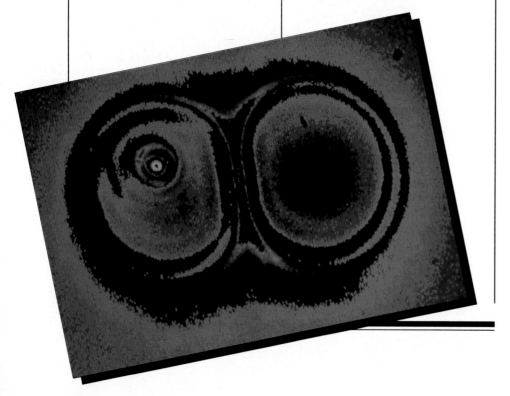

THE EVOLVING BODY

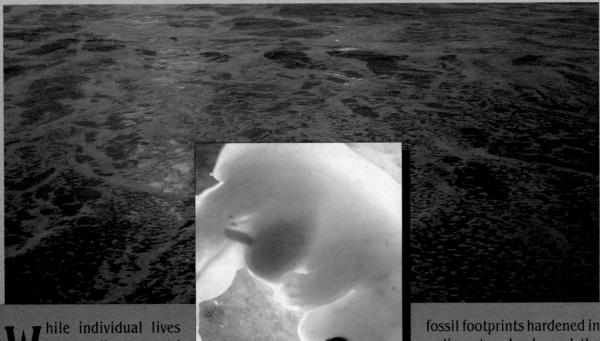

While individual lives are usually measured in hours, days, seasons, and years, human beings can also reckon their existence in terms of the greater spans of time defining their evolution. Thus far, genus *Homo* has barely reached the four-million-year mark on its long journey of gradual change.

Only the most recent human past can be read with anything approaching certainty: Recorded history disappears just a few thousand years from where we stand. And the future, of course, is unknown. Human evolution may continue until the sun dies five billion years or so from now. It may last longer, if we survive that cataclysm—perhaps in some space-roving form. Alternatively, we may evolve no further, the biological progress stopped by some natural or self-inflicted calamity, or by some quirk in the evolutionary process.

Clues to the body's distant past can be found in fragments of prehistoric artifacts,

fossil footprints hardened in sediment and ash, and the physical remains of the early humanlike creatures called hominids. Often discovered accidentally, such evidence shows ancestors both very unlike and very like people today. Our bodies themselves contain clues to our journey through time; they still contain vestiges of biology and behavior acquired in ancient Africa, where humankind is believed to have begun.

How much evolution is going on in us now? No one is certain; the entire span of human history is a blink of an eye compared to the time required for most biological changes to become noticeable. Still, the human body continues to alter. The gradual shrinking of teeth and little toes and fingers testifies to that. Perhaps in time it will be possible to gauge further changes using the genetic information carried within each of our cells. But what form the body ultimately takes must remain, for now, the stuff of imagination.

Footprints in Time

In distance, the trail runs only a hundred feet, a pathway frozen in the hardened volcanic ash of an eastern African gully. But in time, it stretches more than three and a half million years back to the dawn of human history, its preserved set of footprints the earliest known record of a humanlike creature that walked erect.

The trail was discovered in 1976 near the excavations of the famed British paleontologist Mary Leakey in northern Tanzania. Two years earlier, the team digging at the site was about to finish up its work and move on when an accidental find whetted the members' appetites. Feeling playful, a visiting scientist had hurled a lump of dried elephant dung at a colleague, who dodged, tripped, and fell on what proved to be part of an ancient animal trail. The fossil footprints were those of extinct species of giraffe and hyena, preserved in hardened ash. Could there be more prints? And might not some of them belong to a hominid? A long and diligent search of the region eventually turned up what the Leakey team was looking for—the well-defined tracks of a human ancestor.

The signs in the congealed volcanic ash were not hard to read. Evidently, a large hominid, probably a male, had been out front, followed by a smaller, probably female, individual who carefully placed her feet within the prints of

Dental Records

In life, teeth seem all too vulnerable, plagued by erosion and decay. But they are in fact made of such sturdy stuff that they provide one of the most durable records of early humans, a residue in which scientists today can read diet, culture, and even time.

Fossil teeth marked by heavy pitting and scratching suggest a diet of hard fruits, nuts, and seeds. University of Michigan anthropologist C. Loring Brace has hypothesized that the huge front teeth of Neanderthals—humans who died out some 30,000 years ago—may have served as a built-in tool to soften hides or as a vise to hold a wooden shaft, freeing the hands for scraping or carving.

Teeth also have an important chronological story to tell, according to Brace. Comparisons of ancient and contemporary teeth indicate that today's are 45 percent smaller than those of 100,000 years ago. For much of the intervening time, human tooth size declined by about 1 percent every 2,000 years; starting some 10,000 years ago, however, the shrinkage rate doubled. Knowing how rapidly human teeth have been shrinking, scientists can use the sizes of fossil teeth to place them in time.

Although many of these durable remains must be pried from rocky layers of sediments, others are more readily accessible. To the delight of paleontologists, fossil teeth can literally appear overnight, washed from their resting the leader. The presence also of prints of a still smaller hominid suggests that the pair was accompanied by a child.

The tracks lead north, but at one point, the leader seems to have paused, started toward the west, then changed his mind and continued on a northward course. The family of early hominids crossing the African badlands in the depths of time thus left what appears to be a mental as well as a physical record—a very human moment of indecision. □

places by a heavy downpour.

Ancient teeth have also been discovered far from any archaeological digs. Corroborating evidence of the existence of *Australopithecus,* an early human ancestor who lived more than two million years ago, came in 1938, when paleontologist Robert Broom found four of "the most wonderful teeth ever seen in the world's history" in the hands of a fossil-hunting South African schoolboy. The paleontologist begged the boy to take him to the site where the ancient teeth had come from. The boy did so, and from other remains uncovered there, Broom was able to piece together much of a skull that shed light on the life and appearance of this protohuman. □

The Oldest Woman in the World

In 1974, as he searched the scorched wasteland near the Ethiopian town of Hadar for human fossils, American anthropologist Donald Johanson spotted a bit of arm bone beckoning from the gravel of a rain-eroded gulley. Little dreaming of what awaited them, he and his team patiently began digging for further remains. Enough bones would emerge from the gravel for Johanson to be able to piece together about 40 percent of the skeleton of a female australopithecine who had lived more than three million years ago and, assuming that most human of postures, had walked upright.

The skeleton showed that she was short, perhaps only three and a half feet tall. Her head was about the size of a small grapefruit, and, at the ends of long, dangling arms, she had hands similar to those of a modern human.

It was a startling discovery—a breakthrough of the first magnitude. Although fossil fragments had sketched the course of human evolution over the past few million years, no similarly complete hominid skeleton this old had ever been uncovered before. A Neanderthal skeleton previously regarded as being the oldest dated back a mere 75,000 years.

On the night of their find, Johanson and his colleagues sat celebrating, drinking beer, and talking happily while a tape recorder blasted out the Beatles' song *Lucy in the Sky with Diamonds.* Intoxicated by the music and by their

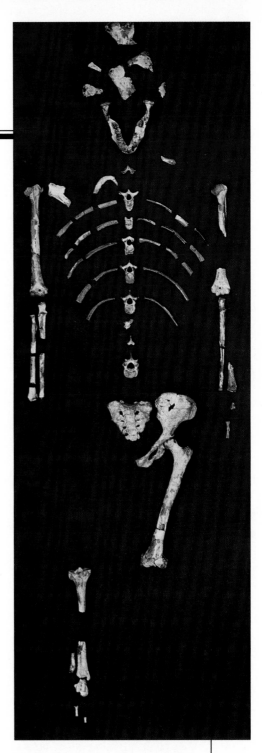

own excitement, the scientists began referring to the little hominid as Lucy, and this was the name that would stick. But their Ethiopian workers had a different name for the oldest woman in the world; they called her Denkenesh, which means, "You are wonderful." □

Confronted by a leopard, a baboon raises its fur to create
an illusion of greater size before making a hasty retreat.
Present-day confrontations between humans still trigger an
ancestral reflex to choose between a fight or flight. But
nowadays the outcome is usually neither—the body suffers
the primitive impulse as increased stress.

Relics of an Animal Past

In its vast inventory of parts, the body includes some that appear to have lost their original function or that seem to serve no purpose whatsoever. Because they often resemble the organs and adaptations of lower and more primitive life forms, these spares are taken to be vestiges of earlier stages of human evolution.

In the human embryo, for example, nonfunctioning gill structures, a two-chambered heart, and three pairs of kidneys evoke a fishlike ancestor. Such evolutionary leftovers are usually transformed in the womb as gestation proceeds. The gill structures, for example, become components of the jaw, ear, and throat. But other anomalies survive whole. A number of embryologists say that the body contains more than a hundred of these, only a few of which are readily apparent.

Best-known is the vermiform (worm-shaped) appendix—the trouble-prone remains of the cecum, an intestinal structure that our plant-eating ancestors needed to digest large quantities of cellulose. Less well-known is the nictitating membrane, which acts as a kind of diving mask in amphibians and an eye-wiping third eyelid in such birds as owls, but exists in us as a small fold of pink tissue in the corner of each eye. Muscles that animals use to aim their ears for better hearing can enable us to wiggle our own ears. Other muscles that pucker our skin into goose bumps are hangovers from a time when, frightened or angry, a hairy ancestor raised his hackles.

Tails, employed to such effect by many creatures, recur in human embryos—at one preliminary stage the tail is one-sixth of the fetus's length. It soon stops growing, turns inward, and by the fourth month fuses into the coccyx, which is located at the base of the spine. Evolution may have erased a visible human tail, but the appendage now and then reappears in some births as a kind of throwback to our early primate past. It is usually boneless, slightly prehensile, and only an inch or so long, although one measuring nine inches has been recorded. □

Are We Eternal Children?

One theory of human evolution holds that we never quite grew up. Certainly, such characteristics as curiosity and innovation can be considered hangovers from childhood. But there may be reason to think that we are childlike in other, more tangible ways.

In 1926, Dutch anatomist Ludwig Bolk noted striking similarities between humans and chimpanzees before and just after birth. The embryonic chimp is hairless, with a flat face, no bony browridges, a rounded head, and small teeth; once born, it continues to bear a striking resemblance to human infants *(below)*. But in the chimp and other primates, these similarities quickly mature into the familiar low-browed, jutting-jaw look of the ape. In humans, however, the same infantile features persist into adulthood, often along with the playful behavior seen in all primate juveniles. In a sense, humans never stop being young.

Besides retaining childlike features and behavioral traits, humans are notoriously slow in their early development compared to various animals. Nine months after conception, human babies come into the world as the most helpless of primate offspring.

According to some researchers, such helplessness suggests that the real period of human gestation is more like twenty-one months. In this view, all babies are born prematurely so that their bodies—their rapidly growing heads especially—can pass through the mother's birth canal. Thus, human newborns may be considered as spending their first year as out-of-the-womb fetuses, vulnerably continuing their gestation even as they begin to learn about the world and its challenges. □

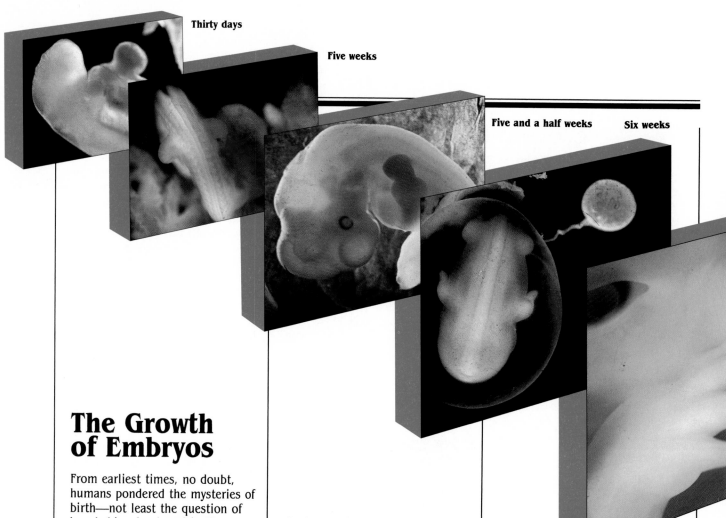

Thirty days

Five weeks

Five and a half weeks   **Six weeks**

The Growth of Embryos

From earliest times, no doubt, humans pondered the mysteries of birth—not least the question of how babies developed in the womb. One possible answer to the developmental puzzle was that the fetus began as a tiny homunculus, or little human, somehow created by the union of man and woman. This belief persisted even into the era of microscopy. The homunculus, some scholars said, was encapsulated in the head of a sperm cell. (The mother's ova, or eggs, were intended merely to nourish the tiny traveler within.) Others insisted that the full-featured little human was latent in the ova but animated by the arrival of sperm.

Not until the mid-eighteenth century did the illogic of preformation become apparent. If all humans were preformed, each tiny homunculus must contain the homunculi of its offspring, and these

the homunculi of theirs, and so on—people within people, like an infinity of nested Russian dolls.

In the nineteenth century, the concept of human evolution and a much more detailed knowledge of the human embryo's stages of development coalesced into something more like scientific theory. Germany's pioneer embryologist Karl Ernst von Baer noted in 1828 that the embryos of many species are similar in their early stages but rapidly diverge thereafter. German biologist Ernst Haeckel took the notion a step further. Stimulated in his thinking by the theory of evolution that was published by the Englishman Charles Darwin in 1859, Haeckel proposed that the process of evolution is repeated by the embryo as it grows in the

womb. Haeckel called his idea the biogenetic law.

Like the idea of the homunculus, however, Haekel's "law" rested on an illusion. Embryos of such vastly different creatures as fish, birds, reptiles, and human beings may seem almost identical, perhaps reflecting a common ancestor at the dawn of life. But this resemblance, it turns out, is superficial at best and persists for only a short interval. Early in their gestation, various embryos rapidly diverge along their separate genetic paths to become the vastly different creatures that emerge at the time of birth. □

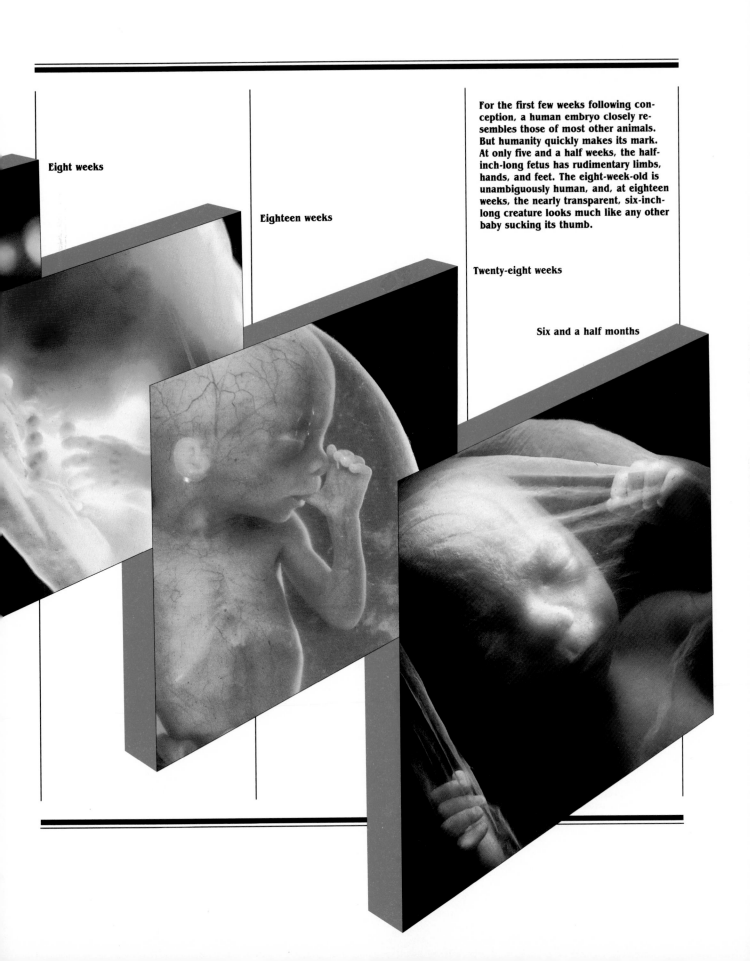

Eight weeks

Eighteen weeks

For the first few weeks following conception, a human embryo closely resembles those of most other animals. But humanity quickly makes its mark. At only five and a half weeks, the half-inch-long fetus has rudimentary limbs, hands, and feet. The eight-week-old is unambiguously human, and, at eighteen weeks, the nearly transparent, six-inch-long creature looks much like any other baby sucking its thumb.

Twenty-eight weeks

Six and a half months

ACKNOWLEDGMENTS

The editors wish to thank these individuals and institutions for their valuable assistance in the preparation of this volume:

François Avril, Départment des Manuscrits, Bibliothèque Nationale, Paris; Robert Bogdan, Syracuse University, New York; Stanley B. Burns, The Burns Archive, New York; Wayne Daniel, Hertzber Circus Collection and Museum, San Antonio, Tex.; Bonnier Fakta, Stockholm; Kathy Falkenstein, Hood College, Frederick, Md.; David W. Frayer, Department of Anthropology, University of Kansas, Lawrence; Thomas H. Goodridge, Silver Spring, Md.; Gunilla Hedesund, Stockholm; Paul Husby, Department of Anaesthesiology, Haukeland Hospital, Bergen, Norway; Bill McCarthy, Circus World Museum, Baraboo, Wis.; Helen Coley Nauts, Science and Medical Communications, Cancer Research Institute, New York; Robert Pelton, The Barnum Museum, Bridgeport, Conn.; Fred D. Pfening III, Circus World Museum, Columbus, Ohio; Peter Pons, Denver Paramedic Division, Denver Department of Health and Hospitals, Colo.; Glee Prahinski, Silver Spring, Md.; Melinda Priestly, Hertzber Circus Collection and Museum, San Antonio, Tex.; Christiane Rajeau, Bibliothèque de l'Ecole Française d'Extrème Orient, Paris; Karla L. Raymond and Warren A. Raymond, Silver Spring, Md.; Franca Maria Ricci, Milan; Lan Swintek, Hertzber Circus Collection and Museum, San Antonio, Tex.; Coleen Tsakopulos, Hertzber Circus Collection and Museum, San Antonio, Tex.; Mary K. Witkowski, Bridgeport Public Library, Conn.; Gretchen Worden, Mútter Museum, College of Physicians of Philadelphia.

PICTURE CREDITS

The sources for the illustrations that appear in this book are listed below. Credits from left to right are separated by semicolons, from top to bottom by dashes.

Cover: © Lennart Nilsson, inset, Dan McCoy/Rainbow. **3:** Dan McCoy/Rainbow. **7:** © Lennart Nilsson, inset, Massimo Listri/KOS, Milan, from *Grande Anatomia del Corpo Umano*, by Paolo Mascagni, published 1823-1831 by Niccolò Capurro, Pisa. **8:** The Mary Evans Picture Library, London. **9:** Lennart Nilsson, *The Body Victorious*, Delacorte Press, New York, 1987; Massimo Listri/KOS, Milan. **10:** Courtesy Centre National de Transfusion Sanguine de Paris. **11:** Michael Holford, Loughton, Essex. **12, 13:** © Lennart Nilsson, *Behold Man*, Little, Brown, Boston, 1973; artwork by Time-Life Books. **14, 15:** Lennart Nilsson, *The Body Victorious*, Delacorte Press, New York, 1987 (2); © Lennart Nilsson, *Behold Man*, Little, Brown, Boston, 1973. **16:** Lennart Nilsson, *Behold Man*, © Little, Brown, Boston, 1973. **17:** Courtesy Dr. Nelson Trujillo. **20, 21:** Photos Lennart Nilsson, © Boehringer Ingelheim International GmbH. **22, 23:** Institut Pasteur, Paris. **24:** © Lennart Nilsson. **25:** Bill Ross/Westlight, inset, Tony Duffy/All Sport Photography, USA. **26, 27:** Mimi Cotter/*People Weekly;* Raymonds of Derby, background from *The Complete Encyclopedia of Illustration*, by J. G. Heck, Park Lane, New York, 1979. **28:** Claude Nuridsary. **29:** Renée Comet. **30, 31:** Steiner Døsvik/Scan-Foto, Oslo; AP/Wide World Photos—Colburn Hudston III/The Forum-Sygma; AP/Wide World Photos. **32, 33:** From *Anomalies and Curiosities of Medicine*, by George M. Gould, M.D., and Walter L. Pyle, M.D., Bell Publishing Company, New York, 1896. **34:** AP/Wide World Photos. **35:** Harold Edgerton 1964, courtesy of Palm Press, Inc. **36:** Tony Duffy/All Sport Photography, USA. **37:** Ronald C. Modra/*Sports Illustrated*. **39:** H. D. Thoreau/ Westlight, inset, Frederick Hill Meserve Collection, National Portrait Gallery, Smithsonian Institution. **40, 41:** From *The Silent Twins*, by Marjorie Wallace, Prentice-Hall Press, New York, 1986. **42:** AP/Wide World Photos. **43:** Mutter Museum, College of Physicians, Philadelphia. **44:** Circus World Museum, Baraboo, Wisconsin. **45:** Frederick Hill Meserve Collection, National Portrait Gallery, Smithsonian Institution. **46:** Courtesy Thomas H. Goodridge. **47:** Courtesy Monte S. Buchsbaum, M.D. **48:** AP/Wide World Photos—Renée Comet, courtesy Glee Prahinski. **51:** Brian Vikander/Westlight, inset, reproduced by kind permission of the President and Council of the Royal College of Surgeons of England. **52:** *Queen Henrietta Maria with Her Dwarf, Sir Anthony van Dyck*, National Gallery of Art, Washington, D.C., Samuel H. Kress Collection. **53:** Reproduced by kind permission of the President and Council of the Royal College of Surgeons of England. **54, 55:** Renée Comet, courtesy Warren and Carla Raymond. **56:** David Frayer, University of Kansas. **57-62:** Renée Comet, courtesy Warren and Carla Raymond. **63:** H. D. Thoreau/ Westlight, inset, Louis Psihoyos/Contact Press Images. **64:** Gianni Dagli Orti, Paris. **65:** From *Sight Restoration after Long-Term Blindness: The Problems and Behavior Patterns of Visual Rehabilitation*, by Alberto Valvo, M.D., American Foundation for the Blind, New York, 1971. **66:** Jean-Loup Charmet, Paris. **67:** © Baron Wolman, 1981/Woodfin Camp—artwork by Time-Life Books. **68:** Background by M. Angelo/ Westlight. **69:** Jean-Loup Charmet, Paris. **70:** © Lennart Nilsson. **71:** Explorer, Paris. **73:** © Lennart Nilsson, *Behold Man*, Little, Brown, Boston, 1973. **74, 75:** Louis Psihoyos/Contact Press Images; Louis Psihoyos, © 1986 National Geographic Society; background artwork by Time-Life Books. **76:** Artwork by Bernt Forsblad. **77-79:** Artwork by Time-Life Books. **80:** Renée Comet, courtesy Lisa Cherkasky. **81:** Brian Vikander/Westlight, inset, Lowell Handler/ Network Photographers, London. **82:** Culver Pictures, Inc. **83:** John Topham Picture Library, Edenbridge, Kent. **84:** Scala/Art Resource, New York; Giraudon, Paris—artwork by Whole Hog Studios from *The Good Cook: Vegetables*, © Time-Life Books, Alexandria, Virginia, 1979. **86:** The Royal Collection, London. **87:** From *The Annotated Alice: Alice's Adventures in Wonderland and Through the Looking Glass*, by Lewis Carroll, Bramhall House, New York, 1960. **88:** Fortean Picture Library, London. **91:** Louis Psihoyos/Matrix. **92:** Lowell Handler/Network Photographers, London. **93:** Antonio Scarpa, courtesy The Bakken, A Museum of Electricity in Life, Minneapolis—courtesy Ron Neafie, parasitologist, and Zan Arnold, photographer, Armed Forces Institute of Pathology, National Museum of Health and Medicine, Washington, D.C., MIS#90-5068. **94:** From *Silhouettes: A Pictorial Archive of Varied Illustrations*, edited by Carol Belanger Grafton, Dover Publications, New York, 1979. **95:** © by Boehringer Ingelheim International GmbH, photo Lennart Nilsson. **97:** Jim Zuckerman/Westlight, inset, Smithsonian Institution (85-18739). **98:** National Institute of Health, Bethesda, Maryland. **99:** Poster by Jules Cheret, courtesy William H. Helfand Collection. **100:** Copied by Steve Tuttle from *The Life of Sir Alexander Fleming, Discoverer of Penicillin*, by André Maurois, Jonathan Cape, London, 1959, courtesy National Library of Medicine, Bethesda, Maryland. **101:** The Bakken, A Museum of Electricity in Life, Minneapolis; National Institute of Health, Bethesda, Maryland—Connecticut Historical Society. **102:** The Bakken, A Museum of Electricity in Life, Minneapolis. **103:** Smithsonian Institution (85-18739). **104:** Artwork by Time-Life Books. **105:** Bettmann Archives. **106, 107:** The Schlesinger Library, Radcliffe College, Cambridge, Massachusetts. **108, 109:** Ed Pierce for *LIFE*. **110:**

BIBLIOGRAPHY

Books

Arey, Leslie Brainerd. *Developmental Anatomy: A Textbook and Laboratory Manual of Embryology* (7th ed.). Philadelphia: W. B. Saunders, 1965.

Atlas of Anatomy. Secaucus, N.J.: Chartwell Books, 1985.

Baker, Susan P., Brian O'Neill, and Ronald S. Karpf. *The Injury Fact Book.* Lexington, Mass.: Lexington Books, 1984.

Balinsky, B. I., and B. C. Fabian. *An Introduction to Embryology* (5th ed.). Philadelphia: Saunders College Publishing, 1981.

Berkow, Robert, M.D. (Ed.). *The Merck Manual of Diagnosis and Therapy* (13th ed.). Rahway, N.J.: Merck Sharp & Dohme Research Laboratories, 1977.

Berlitz, Charles Frambach. *Charles Berlitz's World of Strange Phenomena.* New York: Stonesong Press, 1988.

Berton, Pierre. *The Dionne Years.* New York: W. W. Norton, 1977.

Bodanis, David. *The Body Book: A Fantastic Voyage to the World Within.* Boston: Little, Brown, 1984.

Bogdan, Robert. *Freak Show: Presenting Human Oddities for Amusement and Profit.* Chicago: University of Chicago Press, 1988.

Brooke, John (Ed.). *King George III.* New York: McGraw-Hill, 1972.

Brough, James, et al. *"We Were Five."* New York: Simon & Schuster, 1965.

Brown, Carlton. "Have Midget Will Travel." In *The Circus: Lore and Legend,* edited and compiled by Mildred Sandison Fenner and Wolcott Fenner. Englewood Cliffs, N.J.: Prentice-Hall, 1970.

Campbell, Bernard G. *Human Evolution: An Introduction to Man's Adaptations* (2d ed.). Chicago: Aldine Publishing, 1974.

Campbell, Jeremy. *Winston Churchill's Afternoon Nap.* New York: Simon & Schuster, 1986.

Carlson, Bruce M., M.D. *Patten's Foundations of Embryology* (4th ed.). New York: McGraw-

Hill, 1981.

Caselli, Giovanni. *The Human Body and How It Works.* New York: Grosset & Dunlap, 1987.

Cassill, Kay. *Twins: Nature's Amazing Mystery.* New York: Atheneum, 1982.

Caufield, Catherine. *Multiple Exposures: Chronicles of the Radiation Age.* New York: Harper & Row, 1989.

Cytowic, Richard E. *Synesthesia: A Union of the Senses.* New York: Springer-Verlag, 1989.

Davison, Gerald C., and John M. Neale. *Abnormal Psychology: An Experimental Clinical Approach* (3d ed.). New York: John Wiley & Sons, 1982.

Dewey, Edward R., and Og Mandino. *Cycles: The Mysterious Forces That Trigger Events.* New York: Hawthorn Books, 1971.

Dobzhansky, Theodosius. *Evolution, Genetics, and Man.* New York: John Wiley & Sons, 1955.

Drimmer, Frederick:
Born Different: Amazing Stories of Very Special People. New York: Atheneum, 1988.
Very Special People: The Struggles, Loves, and Triumphs of Human Oddities. New York: Amjon Publishers, 1973.

Edwards, Frank:
Strange People. New York: Lyle Stuart, 1961.
Strangest of All. Secaucus, N.J.: Citadel Press, 1984.

Fenton, Carroll Lane. *Our Living World.* Garden City, N.Y.: Doubleday, Doran, 1945.

Fiedler, Leslie. *Freaks: Myths and Images of the Secret Self.* New York: Simon & Schuster, 1978.

Gallagher, Richard. *Diseases That Plague Modern Man.* Dobbs Ferry, N.Y.: Oceana, 1969.

Galton, Lawrence. *1,001 Health Tips.* New York: Simon & Schuster, 1984.

Gates, Reginald Ruggles. *Human Genetics* (Vol. 1). New York: Macmillan, 1946.

Gilbert, Scott F. *Developmental Biology* (2d ed.). Sunderland, Mass.: Sinauer Associates, 1988.

Gilling, Dick, and Robin Brightwell. *The Human Brain.* New York: Facts On File Publications,

1982.

Gould, George M., M.D., and Walter L. Pyle, M.D. *Anomalies and Curiosities of Medicine.* New York: Bell, 1896.

Gould, Rupert T. *Enigmas: Another Book of Unexplained Facts.* New Hyde Park, N.Y.: University Books, 1965.

Gould, Stephen Jay. *Ontogeny and Phylogeny.* Cambridge, Mass.: Belknap Press, 1977.

Guinness Book of World Records 1989. New York: Sterling, 1988.

Hand, Wayland D. *Magical Medicine.* Berkeley: University of California Press, 1980.

Hechtlinger, Adelaide. *The Great Patent Medicine Era.* New York: Grosset & Dunlap, 1972.

Howard, J. Keir, and Frank H. Tyrer (Eds.). *Textbook of Occupational Medicine.* Edinburgh: Churchill Livingstone, 1987.

Jackson, Gordon, and Philip Whitfield. *Digestion: Fueling the System.* New York: Torstar Books, 1984.

Jastrow, Joseph. *Fact and Fable in Psychology.* Freeport, N.Y.: Books For Libraries Press, 1971.

Johanson, Donald C., and Maitland A. Edey. *Lucy: The Beginnings of Humankind.* New York: Simon & Schuster, 1981.

Jones, Howard W., Jr., M.D., and Georgeanna Seegar Jones, M.D. *Novak's Textbook of Gynecology* (10th ed.). Baltimore: Williams & Wilkins, 1981.

Karp, Gerald, and N. J. Berrill. *Development* (2d ed.). New York: McGraw-Hill, 1981.

Kendig, Frank, and Richard Hutton. *Life-Spans: Or How Long Things Last.* New York: Holt, Rinehart & Winston, 1979.

King, Francis. *Wisdom from Afar.* London: Danbury Press, 1975.

Klaus, Marshall, M.D., and Phyllis H. Klaus (Eds.). *The Amazing Newborn.* Reading, Mass.: Addison-Wesley, 1985.

Knight, Bernard, M.D. *Discovering the Human Body: How Pioneers of Medicine Solved the Mysteries of the Body's Structure and Func-*

tion. New York: Lippincott & Crowell, 1980.

Krupp, Marcus A., M.D., Milton J. Chatton, M.D., and Lawrence M. Tierney, Jr., M.D. *Current Medical Diagnosis & Treatment 1986*. Los Altos, Calif.: Lange Medical Publications, 1986.

Kunz, Jeffrey R. M., M.D. (Ed.). *The American Medical Association Family Medical Guide*. New York: Random House, 1982.

Leakey, Richard E. *Human Origins*. New York: E. P. Dutton, 1982.

Lifson, Robert. *Enter the Sideshow*. Bala Cynwyd, Pa.: Mason, 1983.

Lindsey, Arthur Ward. *Principles of Organic Evolution*. St. Louis: C. V. Mosby, 1952.

Locke, Steven, M.D., and Douglas Colligan. *The Healer Within*. New York: E. P. Dutton, 1986.

Louis, David. *2201 Fascinating Facts*. New York: Greenwich House, 1983.

Lyons, Albert S., M.D., and R. Joseph Petrucelli II, M.D. *Medicine: An Illustrated History*. New York: Harry N. Abrams, 1978.

McAleer, Neil. *The Body Almanac: Mind-Boggling Facts about Today's Human Body and High-Tech Medicine*. Garden City, N.Y.: Doubleday, 1985.

McCall, Robert B. *Infants*. Cambridge, Mass.: Harvard University Press, 1979.

McCutcheon, Marc. *The Compass in Your Nose and Other Astonishing Facts about Humans*. Los Angeles: Jeremy P. Tarcher, 1989.

McEvoy, J. P. "The Siamese Twins." In *The Circus: Lore and Legend*, edited and compiled by Mildred Sandison Fenner and Wolcott Fenner. Englewood Cliffs, N.J.: Prentice-Hall, 1970.

Maddox, Sam. *Spinal Network*. Boulder, Colo.: Spinal Network, 1987.

Marnham, Patrick. *Lourdes: A Modern Pilgrimage*. New York: Coward, McCann & Geoghegan, 1981.

Maurois, André. *The Life of Sir Alexander Fleming*. Translated by Gerard Hopkins. London: Jonathan Cape, 1959.

Melzack, Ronald, and Patrick D. Wall. *The Challenge of Pain*. New York: Basic Books, 1973.

Merrell, David J. *Evolution and Genetics*. New York: Holt, Rinehart & Winston, 1962.

Miller, Jonathan. *The Body in Question*. New York: Random House, 1978.

Mizel, Steven B., and Peter Jaret. *In Self-Defense*. New York: Harcourt Brace Jovanovich, 1985.

Monestier, Martin. *Human Oddities*. Translated by Robert Campbell. Secaucus, N.J.: Citadel Press, 1987.

Moody, Paul Amos. *Introduction to Evolution* (2d ed.). New York: Harper & Brothers, 1962.

Murphy, Wendy, and the Editors of Time-Life Books. *Touch, Taste, Smell, Sight and Hearing* (Library of Health series). Alexandria, Va.: Time-Life Books, 1982.

Mysteries of the Unexplained. Pleasantville, N.Y.: Reader's Digest Association, 1982.

Nelsen, Olin E. *Comparative Embryology of the Vertebrates*. New York: McGraw-Hill, 1953.

Newby, Hayes A. *Audiology* (4th ed.). Englewood Cliffs, N.J.: Prentice-Hall, 1979.

Nilsson, Lennart, and Jan Lindberg:
Behold Man. Translated by Ilona Munck. Boston: Little, Brown, 1973.
The Body Victorious. Translated by Clare James. New York: Delacorte Press, 1987.

Nolen, William A., M.D. *Healing: A Doctor in Search of a Miracle*. New York: Random House, 1974.

Nourse, Alan E., and the Editors of Time-Life Books. *The Body* (Life Science Library series). New York: Time-Life Books, 1971.

Novotny, Ann, and Carter Smith (Eds.). *Images of Healing*. New York: Macmillan, 1980.

Pritchard, Jack A., Paul C. MacDonald, and Normand F. Gant (Eds.). *Williams Obstetrics* (17th ed.). Norwalk, Conn.: Appleton-Century-Crofts, 1985.

Psychology Today: An Introduction. Del Mar, Calif.: CRM Books, 1970.

Ravitch, Mark M., M.D., et al. (Eds.). *Pediatric Surgery*. Chicago: Year Book Medical Publishers, 1979.

Reader, John. *Missing Links: The Hunt for Earliest Man*. Boston: Little, Brown, 1981.

Reader's Digest Book of Facts. Pleasantville, N.Y.: Reader's Digest Association, 1987.

Restak, Richard M., M.D.:
The Brain. Toronto: Bantam Books, 1984.
The Mind. Toronto: Bantam Books, 1988.

Rosenfeld, Albert. *Pro Longevity*. New York: Alfred A. Knopf, 1976.

Rosenthal, David (Ed.). *The Genain Quadruplets*. New York: Basic Books, 1963.

Roth, Hy, and Robert Cromie. *The Little People*. New York: Everest House, 1980.

Roueché, Berton. *The Orange Man and Other Narratives of Medical Detection*. Boston: Little, Brown, 1971.

Sacks, Oliver. *The Man Who Mistook His Wife for a Hat and Other Clinical Tales*. New York: Harper & Row, 1987.

Sanderson, Ivan T. *Investigating the Unexplained*. Englewood Cliffs, N.J.: Prentice-Hall, 1972.

Seizer, Richard. *Mortal Lessons: Notes on the Art of Surgery*. New York: Simon & Schuster, 1976.

Siegel, Bernie S., M.D. *Love, Medicine & Miracles*. New York: Harper & Row, 1986.

Sittig, Marshall. *Handbook of Toxic and Hazardous Chemicals and Carcinogens* (2d ed.). Park Ridge, N.J.: Noyes Publications, 1985.

Smith, Anthony:
The Body. New York: Viking Press, 1986.
The Mind. New York: Viking Press, 1984.

Stage, Sarah. *Female Complaints*. New York: W. W. Norton, 1979.

Strange Stories, Amazing Facts. Pleasantville, N.Y.: Reader's Digest Association, 1976.

Toomey, Bill, and Barry King. *The Olympic Challenge*. Reston, Va.: Reston Publishing, 1984.

Trevor-Roper, Patrick. *The World through Blunted Sight*. Indianapolis: Bobbs-Merrill, 1970.

Waddington, C. H. *Principles of Embryology*. London: George Allen & Unwin, 1960.

Wallace, Irving, David Wallechinsky, and Amy Wallace. *Significa*. New York: E. P. Dutton, 1983.

Wallace, Marjorie. *The Silent Twins*. New York: Prentice-Hall Press, 1986.

Walraven, Michael Guiford, and Hiram E. Fitzgerald (Eds.). *Psychology, 89/90* (19th ed.) (Annual Editions series). Guilford, Conn.: Dushkin Publishing, 1989.

Walsh, James J., M.D. *Cures: The Story of the Cures That Fail*. New York: D. Appleton, 1923.

Ward, Ritchie R. *The Living Clocks*. New York: Alfred A. Knopf, 1977.

Watson, Peter. *Twins: An Uncanny Relationship?* Chicago: Contemporary Books, 1981.

Wilentz, Joan Steen. *The Senses of Man*. New York: Thomas Y. Crowell, 1968.

Winfree, Arthur T. *The Timing of Biological Clocks*. New York: Scientific American Books, 1987.

Wolf, Fred Alan. *The Body Quantum: The New Physics of Body, Mind and Health*. New York: Macmillan, 1986.

Yahr, Melvin D., M.D. (Ed.). *Transactions of the American Neurological Association, 1964* (Vol. 89). New York: Springer Publishing, 1964.

Young, James Harvey:
The Medical Messiahs. Princeton, N.J.: Princeton University Press, 1967.
The Toadstool Millionaires. Princeton, N.J.: Princeton University Press, 1961.

Periodicals

Abramson, Pamela. "Mysteries of the Sipstakes." *Newsweek*, September 16, 1985.

"An Account of the Death of the Countess Cornelia Baudi of Cesena." *Annual Register*, 1763, pp. 91-95.

Ahrens, Frank. "Flesh-Eating Maggots Save Teen-Ager's Leg." *Montgomery Journal*, August 3, 1989.

Alder, Jerry. "The Merciless Man of Wine." *Newsweek*, December 14, 1987.

"All about Twins." *Newsweek*, November 23, 1987.

"Alton Giant." *Time*, March 1, 1937.

Alvarado, Carlos S. "Observations of Luminous Phenomena around the Human Body: A Review." *Journal of the Society for Psychical Research*, 1987, Vol. 54, p. 38.

Amato, Ivan:

"Aroma Tech: Scents and Sensibility." *Washington Post*, May 22, 1988.

"Future Chomp: The Science of Flavor." *Washington Post*, July 7, 1988.

Ammons, Carol H., et al. " 'Facial Vision': The Perception of Obstacles Out of Doors by Blindfolded and Blindfolded Deafened Subjects." *American Journal of Psychology*, 1953, Vol. 66, pp. 519-553.

Ausubel, Ken. "The Troubling Case of Harry Hoxsey." *NewAge Journal*, July-August 1988.

Bailey, George, M.D. "Toxic Epidermal Necrolysis." *JAMA*, March 22, 1965.

Berk, Marc L., Alan C. Monheit, and Michael M. Hagan. "How the U.S. Spent Its Health Care Dollar: 1929-1980." *Health Affairs*, Fall 1988.

Boldt, David. "The Strange Cases of Dr. Krogman." *TODAY/The Philadelphia Inquirer*, June 15, 1975.

Bower, B. "The Left Hand of Math and Verbal Talent." *Science News*, 1985, Vol. 127, p. 263.

Brody, Jane E.:

"For Artists and Musicians, Creativity Can Mean Illness and Injury." *New York Times*, October 17, 1989.

"Sense of Smell Proves to Be Surprisingly Subtle." *New York Times*, February 2, 1983.

Brunn, Ruth Dowling, M.D. "Gilles de la Tourette's Syndrome: An Overview of Clinical Experience." *Journal of the American Academy of Child Psychiatry*, 1984, Vol. 23, no. 2.

Buchsbaum, Monte S. "The Genain Quadruplets." *Psychology Today*, August 1984.

Buffington, Perry W. "Making (Body) Sense." *Sky*, November 1983.

Cassill, Kay. "The Phantom Twin." *Health*, April 1983.

Cool, Lisa Collier. "Fatigue." *McCall's*, December 1987.

"Couple Sharing 70 Years Die 23 Hours Apart." *Pursuit*, 1987, Vol. 20, p. 81.

Cunningham, Susan. "Is It One Clock, or Several, That Cycle Us through Life?" *APA Monitor*, September 1985.

Curtis, H. A., et al. "Mercury as a Health Hazard." *Archives of Disease in Childhood*, March 1987.

Davis, Joel. "Smell Lab." *Omni*, January 1984.

Demak, Richard. "Fighting the Enemy Within." *Sports Illustrated*, June 22, 1987.

Dickey, Marilyn. "Lifting Depression." *Washingtonian*, March 1988.

Durden-Smith, Jo, and Diane DeSimone. "Hidden Threads of Illness." *Science Digest*, January 1984.

Ecenbarger, William. "Just for the Smell of It." *Washington Post*, July 28, 1987.

Ehrenkranz, Joel R. L. "A Gland for All Seasons." *Natural History*, March 1988.

Elson, John. "The Man with a Paragon Palate." *Time*, December 14, 1987.

"Fallible Doctors." *The Economist*, December 17, 1988.

"Fighting the Winter Blues with Bright Light." *Psychology Today*, January-February 1989.

Friedman, Emily. "Clinical Conservatism Keeps U.K. Angiography Rate Low." *Medical World News*, September 11, 1989.

Frisch, Joan Vandiver. "The Sound in the Silence." *NOAA*, October 1973.

Garelik, Glenn. "Exorcising a Damnable Disease." *Discover*, December 1986.

Gibson, Eleanor J., and Richard D. Walk. "The 'Visual Cliff.' " *Scientific American*, April 1960.

Goleman, Daniel. "Feeling Sleepy? An Urge to Nap is Built In." *New York Times*, September 12, 1989.

Gould, Judy. "What Happens When Three Young Men Find Out They're Triplets? It's Not as Simple as 1-2-3." *People*, October 10, 1980.

Grist, Liz. "Why Most People Are Right-Handed." *New Scientist*, August 16, 1984.

Halliday, D. J. X., et al. "Human Combustion" (letter to the editor). *New Scientist*, May 29, 1986.

Halpern, Diane F., and Stanley Coren. "Do Right-Handers Live Longer?" *Nature*, May 19, 1988.

Heymer, John:

"A Burnt-Out Case?" *New Scientist*, May 19, 1988.

"A Case of Spontaneous Human Combustion?" *New Scientist*, May 15, 1986.

Holden, Constance:

"Genes and Behavior: A Twin Legacy." *Psychology Today*, September 1987.

"The Genetics of Personality." *Science*, August 7, 1987.

Huey, John. "Let Them Eat Dirt." *SouthPoint*, October 1989.

Jaret, Peter. "Our Immune System: The Wars Within." *National Geographic*, June 1986.

Juan, Stephen. "Combustion Theory Goes Up in Smoke." *Sydney Morning Herald*, March 8, 1989.

Kernan, Michael. "Remembrances of Those Who Fell from the Heights." *Smithsonian*, July 1981.

Lebow, Howard A. "Have You Heard about This One?" *Cycles*, 1986, Vol. 37, p. 191.

"Lend Me a Hand." *Washington Times*, September 20, 1989.

Leo, John. "The Hidden Power of Body Odors." *Time*, December 1, 1986.

Litsky, Frank. "How Fast, How Far? No Limits in Sight." *New York Times*, June 16, 1986.

Lockwood, Alan H., M.D. "Medical Problems of Musicians." *New England Journal of Medicine*, January 26, 1989.

Long, Michael E. "What Is This Thing Called Sleep?" *National Geographic*, December 1987.

Lubaroff, Saul. "Tourette's Sufferers Can't Help but Say What's on Their Minds—Even When It Hurts." *People*, March 23, 1987.

Lyali, Sarah. "Alexander M. Lewyt Dead at 79." *New York Times*, March 21, 1988.

McIver, Beale. "Sergeant Alkemade: The Great Escape." *The Unexplained Mysteries of Mind Space & Time*, Vol. 7, no. 75.

Maluf, N. S. R. "History of Blood Transfusion." *Journal of the History of Medicine and Allied Sciences*, January 1954.

"Medical Curiosities: Holes in Heads." *Fortean Times*, Autumn 1982.

Miller, Julie Ann. "Clockwork in the Brain." *BioScience*, February 1989.

Mirsky, Allan F., and Olive W. Quinn. "The Genain Quadruplets." *Schizophrenia Bulletin*, 1988, Vol. 14, no. 4.

Monmaney, Terence:

"Are We Led by the Nose?" *Discover*, September 1987.

"The Chemistry between People." *Newsweek*, January 12, 1987.

Musto, David F. "America's First Cocaine Epidemic." *The Wilson Quarterly*, Summer 1989.

Nickell, Joe, and John F. Fischer. "Did Jack Angel Survive Spontaneous Combustion?" *Fate*, May 1989.

Oakes, Larry. "Carpenter Recovers after Hitting a Nail Right in the Head." *Washington Times*, October 12, 1989.

"On the Spontaneous Combustion of the Human Body." *Edinburgh New Philosophical Journal*, 1828, Vol. 5, pp. 164-169.

Osborne, John P., P. Munson, and D. Burman. "Huntington's Chorea." *Archives of Disease in Childhood*, February 1982.

O'Shea, J. G. "The Death of Paganini." *Journal of the Royal College of Physicians of London*, April 1988.

Parachini, Allan. "A Baseball Player's Life Linked to His Stats." *Los Angeles Times*, May 12, 1988.

Passouant, Pierre. "Doctor Gelineau (1828-1906): Narcolepsy Centennial." *Sleep*, 1981,

Vol. 4, no. 3.

Pechter, Edward A., M.D., and Ronald A. Sherman. "Maggot Therapy: The Surgical Metamorphosis." *Plastic and Reconstructive Surgery,* October 1983.

Rose, Richard J., et al. "Shared Genes, Shared Experiences, and Similarity of Personality: Data from 14,288 Adult Finnish Co-Twins." *Journal of Personality and Social Psychology,* January 1988.

Rosenbaum, Ron, and Susan Edmiston. "Dead Ringers." *Esquire,* March 1976.

Sachs, Oliver. "Being Moved by the Spirit." *Sunday Times Magazine,* September 25, 1988.

Sacks, Oliver:
"The Divine Curse." *Life,* September 25, 1988.
"Rude Awakening." *Discover,* February 1988.

Salzman, Mark. "WUSHU: Meditation in Motion." *New York Times,* March 29, 1987.

Shulman, Seth. "Scientists Trying to Sniff Out Secrets of the Sense of Smell." *Los Angeles Times,* July 25, 1988.

"Sick Health Services." *The Economist,* July 16, 1988.

Siegel, Jacob S., and Jeffrey S. Passel. "New Estimates of the Number of Centenarians in the United States." *Journal of the American Statistical Association,* September 1976.

"Spontaneous Human Combustion?" *British Medical Journal,* 1938, Part 1, p. 1106.

"The State of Our 50 (plus D.C.)." *Life,* Vol. 12, no. 12.

Stein, Kathleen. "Scent of Sex." *Omni,* April 1980.

Stewart, Jonathan T., M.D. "Huntington's Disease." *American Family Physician,* May

1988.

Stocking, Barbara. "Medical Technology in the United Kingdom." *International Journal of Technology Assessment in Health Care,* 1988, Vol. 4, pp. 171-183.

Stockwell, G. Archie. "Spontaneous Human Combustion." *Scientific American Supplement,* 1889, Vol. 28.

Swedo, Susan E., M.D., et al. "A Double-Blind Comparison of Clomipramine and Desipramine in the Treatment of Trichotillomania (Hair Pulling)." *New England Journal of Medicine,* August 24, 1989.

Tart, Charles T. "Concerning the Scientific Study of the Human Aura." *Journal of the Society for Psychical Research,* March 1972.

"Teeth Are Shrinking." *New Scientist,* November 26, 1987.

"Testing Human Limits." *U.S. News & World Report,* September 19, 1988.

Thatcher, Gary. " 'Anthrax' Island: Where Life Imitates Fiction." *Christian Science Monitor,* December 15, 1988.

Thompson, Marcia J., and David W. Harsha. "Our Rhythms Still Follow the African Sun." *Psychology Today,* January 1984.

Thurston, Gavin. "Spontaneous Human Combustion." *British Medical Journal,* 1938, Part 1, p. 1340.

Trimble, Michael. "Psychopathology and Movement Disorders: A New Perspective on the Gilles de la Tourette Syndrome." *Journal of Neurology, Neurosurgery, and Psychiatry,* Special Supplement, June 1989, pp. 90-95.

Waters, Tom. "Sinister Stats." *Discover,* April 1989.

"Wed When the West was Wild, Ernie and Maud Scott Celebrate 80 Years of Staying Hitched." *People,* June 26, 1989.

Weiss, Rick. "The Viral Advantage." *Science News,* September 23, 1989.

"What's It Like to Fall 3,600 Feet with Failed Parachutes? 'My Life Really Did Pass Before Me.' " *People,* Oct 10, 1977.

"Where Schmaltz Is Standard Fare." *Insight,* October 16, 1989.

Williams, Roger. "Twin Dreams." *Science Digest,* November 1982.

Wittner, Murray, M.D., et al. "Eustrongylidiasis: A Parasitic Infection Acquired by Eating Sushi." *New England Journal of Medicine,* April 27, 1989.

"Woman Bursts into Flames on the Street." *Buffalo Evening News,* August 6, 1982.

Worchel, Philip, et al. "The Perception of Obstacles by the Blind." *Journal of Experimental Psychology,* 1950, Vol. 40, pp. 746-751.

Worchel, Philip, and Karl M. Dallenbach. " 'Facial Vision:' Perception of Obstacles by the Deaf-Blind." *American Journal of Psychology,* 1947, Vol. 60, pp. 502-553.

Other Sources

"Facts Concerning Johnny Eck: The Only Living Half-Boy." Promotional pamphlet, 1930s.

"New Identical Twins Study." Transcript of ABC News Nightline Show no. 2181, October 2, 1989.

O'Regan, Brendan. "Healing, Remission and Miracle Cures." Special report of a lecture on December 5, 1986. Washington, D.C.: Institute of Noetic Sciences, May 1987.

Index

Time-Life Books Inc.
is a wholly owned subsidiary of
THE TIME INC. BOOK COMPANY

President and Chief Executive Officer:
Kelso F. Sutton
President, Time Inc. Books Direct:
Christopher T. Linen

TIME-LIFE BOOKS INC.

EDITOR: George Constable
Executive Editor: Ellen Phillips
Director of Design: Louis Klein
Director of Editorial Resources: Phyllis K. Wise
Director of Photography and Research:
John Conrad Weiser

PRESIDENT: John M. Fahey, Jr.
Senior Vice Presidents: Robert M. DeSena,
Paul R. Stewart, Curtis G. Viebranz, Joseph J. Ward
Vice Presidents: Stephen L. Bair, Bonita L.
Boezeman, Mary P. Donohoe, Stephen L. Goldstein,
Juanita T. James, Andrew P. Kaplan, Trevor Lunn,
Susan J. Maruyama, Robert H. Smith
New Product Development: Yuri Okuda,
Donia Ann Steele
Supervisor of Quality Control: James King

PUBLISHER: Joseph J. Ward

Editorial Operations
Copy Chief: Diane Ullius
Production: Celia Beattie
Library: Louise D. Forstall
Computer Composition: Gordon E. Buck
(Manager), Deborah G. Tait, Monika D. Thayer,
Janet Barnes Syring, Lillian Daniels

**Library of Congress
Cataloging in Publication Data**
Mysteries of the human body / by the editors of
Time-Life Books.
p. cm. (Library of curious and unusual facts).
Bibliography: p.
Includes index.
ISBN 0-8094-7679-7
ISBN 0-8094-7680-0 (lib. bdg.)
1. Human physiology—Popular works.
2. Body, Human—Popular works.
I.Time-Life Books. II. Series.
QP38.M87 1990
612—dc20   90-30146 CIP

LIBRARY OF CURIOUS AND UNUSUAL FACTS

SERIES DIRECTOR: Russell B. Adams, Jr.
Series Administrator: Elise Dawn Ritter
Designer: Susan K. White
Associate Editor: Sally Collins (pictures)

Editorial Staff for
Mysteries of the Human Body
Text Editors: Carl A. Posey (principal),
Laura Foreman
Researchers: Maureen McHugh (principal),
Sydney J. Baily, M. Tucker Jones, Robert H.
Wooldridge, Jr.
Assistant Designer: Alan Pitts
Copy Coordinators: Jarelle S. Stein (principal),
Anne Farr, Darcie Conner Johnston
Picture Coordinator: Leanne G. Miller
Editorial Assistant: Terry Ann Paredes

Special Contributors: Lesley Coleman, Christine
Hinze (London research); Joe Alper, Bruce Fried-
land, Carollyn James, Peter Kaufman, Debra Kent,
Gina Maranto, Peter Pocock, George Russell,
William G. Shepherd, Jr., Michael Woods (text);
Philip Murphy, Christopher J. Napierala, Douglas C.
Nelson, Kathryn Pfeifer, Cornelia M. Piper, Evelyn
Prettyman, Norma Shaw, Debra Diamond Smit
(research); Victoria Agee (index)

Correspondents: Elisabeth Kraemer-Singh (Bonn),
Christina Lieberman (New York), Maria Vincenza
Aloisi (Paris), Ann Natanson (Rome).
Valuable assistance was also provided by Elizabeth
Brown (New York); Dag Christensen (Oslo); Ann Wise
(Rome); Mary Johnson (Stockholm); Dick Berry,
Mieko Ikeda (Tokyo); Traudl Lessing (Vienna).

The Consultants:
William R. Corliss, the general consultant for the
series, is a physicist-turned-writer who has spent the
last twenty-five years compiling collections of
anomalies in the fields of geophysics, geology, ar-
chaeology, astronomy, biology, and psychology. He
has written about science and technology for NASA,
the National Science Foundation, and the Energy
Research and Development Administration (among
others). Mr. Corliss is also the author of more than
thirty books on scientific mysteries, including *Mysterious Universe, The Unfathomed Mind,* and *Handbook of Unusual Natural Phenomena.*

Bruce H. Dobkin, M.D., is a professor of neurology
at the UCLA School of Medicine. He is the author of
*Brain Matters: Stories of a Neurologist and His Pa-
tients.*

Steven B. Mizel, Ph.D., is chairman of the Department of Microbiology and Immunology at the Bow-
man Gray School of Medicine at Wake Forest University in Winston-Salem, North Carolina. He is the
coauthor of *In Self-Defense,* a popular account of
the immune system.

Marcello Truzzi, a professor of sociology at Eastern
Michigan University, is also director of the Center
for Scientific Anomalies Research (CSAR) and editor
of its journal, *Zetetic Scholar.*

Other Publications:

AMERICAN COUNTRY
VOYAGE THROUGH THE UNIVERSE
THE THIRD REICH
THE TIME-LIFE GARDENER'S GUIDE
MYSTERIES OF THE UNKNOWN
TIME FRAME
FIX IT YOURSELF
FITNESS, HEALTH & NUTRITION
SUCCESSFUL PARENTING
HEALTHY HOME COOKING
UNDERSTANDING COMPUTERS
LIBRARY OF NATIONS
THE ENCHANTED WORLD
THE KODAK LIBRARY OF CREATIVE PHOTOGRAPHY
GREAT MEALS IN MINUTES
THE CIVIL WAR
PLANET EARTH
COLLECTOR'S LIBRARY OF THE CIVIL WAR
THE EPIC OF FLIGHT
THE GOOD COOK
WORLD WAR II
HOME REPAIR AND IMPROVEMENT
THE OLD WEST

*For information on and a full description of any of
the Time-Life Books series listed above, please call
1-800-621-7026 or write:*
Reader Information
Time-Life Customer Service
P.O. Box C-32068
Richmond, Virginia 23261-2068

This volume is one in a series that explores
astounding but surprisingly true events in history,
science, nature, and human conduct.

Time-Life Books Inc. offers a wide range of fine re-
cordings, including a *Rock 'n' Roll Era* series. For
subscription information, call 1-800-621-7026 or
write Time-Life Music, P.O. Box C-32068, Richmond,
Virginia 23261-2068.

The information in this book cannot and should not
replace the medical advice of a physician. In the
event of illness, you should consult a qualified
health professional.